LONDON MATHEMATICAL SOCIETY LECTURE NOTI

Managing Editor: Professor I.M. James,
Mathematical Institute, 24-29 St Giles,O:

1. General cohomology theory and K-theor
4. Algebraic topology, J.F.ADAMS
5. Commutative algebra, J.T.KNIGHT
8. Integration and harmonic analysis on compact groups, R.E.EDWARDS
9. Elliptic functions and elliptic curves, P.DU VAL
10. Numerical ranges II, F.F.BONSALL & J.DUNCAN
11. New developments in topology, G.SEGAL (ed.)
12. Symposium on complex analysis, Canterbury, 1973, J.CLUNIE & W.K.HAYMAN (eds.)
13. Combinatorics: Proceedings of the British Combinatorial Conference 1973, T.P.McDONOUGH & V.C.MAVRON (eds.)
15. An introduction to topological groups, P.J.HIGGINS
16. Topics in finite groups, T.M.GAGEN
17. Differential germs and catastrophes, Th.BROCKER & L.LANDER
18. A geometric approach to homology theory, S.BUONCRISTIANO, C.P. ROURKE & B.J.SANDERSON
20. Sheaf theory, B.R.TENNISON
21. Automatic continuity of linear operators, A.M.SINCLAIR
23. Parallelisms of complete designs, P.J.CAMERON
24. The topology of Stiefel manifolds, I.M.JAMES
25. Lie groups and compact groups, J.F.PRICE
26. Transformation groups: Proceedings of the conference in the University of Newcastle-upon-Tyne, August 1976, C.KOSNIOWSKI
27. Skew field constructions, P.M.COHN
28. Brownian motion, Hardy spaces and bounded mean oscillations, K.E.PETERSEN
29. Pontryagin duality and the structure of locally compact Abelian groups, S.A.MORRIS
30. Interaction models, N.L.BIGGS
31. Continuous crossed products and type III von Neumann algebras, A.VAN DAELE
32. Uniform algebras and Jensen measures, T.W.GAMELIN
33. Permutation groups and combinatorial structures, N.L.BIGGS & A.T.WHITE
34. Representation theory of Lie groups, M.F. ATIYAH et al.
35. Trace ideals and their applications, B.SIMON
36. Homological group theory, C.T.C.WALL (ed.)
37. Partially ordered rings and semi-algebraic geometry, G.W.BRUMFIEL
38. Surveys in combinatorics, B.BOLLOBAS (ed.)
39. Affine sets and affine groups, D.G.NORTHCOTT
40. Introduction to Hp spaces, P.J.KOOSIS
41. Theory and applications of Hopf bifurcation, B.D.HASSARD, N.D.KAZARINOFF & Y-H.WAN
42. Topics in the theory of group presentations, D.L.JOHNSON
43. Graphs, codes and designs, P.J.CAMERON & J.H.VAN LINT
44. Z/2-homotopy theory, M.C.CRABB
45. Recursion theory: its generalisations and applications, F.R.DRAKE & S.S.WAINER (eds.)
46. p-adic analysis: a short course on recent work, N.KOBLITZ
47. Coding the Universe, A.BELLER, R.JENSEN & P.WELCH
48. Low-dimensional topology, R.BROWN & T.L.THICKSTUN (eds.)

49. Finite geometries and designs, P.CAMERON, J.W.P.HIRSCHFELD & D.R.HUGHES (eds.)
50. Commutator calculus and groups of homotopy classes, H.J.BAUES
51. Synthetic differential geometry, A.KOCK
52. Combinatorics, H.N.V.TEMPERLEY (ed.)
53. Singularity theory, V.I.ARNOLD
54. Markov processes and related problems of analysis, E.B.DYNKIN
55. Ordered permutation groups, A.M.W.GLASS
56. Journées arithmétiques 1980, J.V.ARMITAGE (ed.)
57. Techniques of geometric topology, R.A.FENN
58. Singularities of smooth functions and maps, J.MARTINET
59. Applicable differential geometry, M.CRAMPIN & F.A.E.PIRANI
60. Integrable systems, S.P.NOVIKOV et al.
61. The core model, A.DODD
62. Economics for mathematicians, J.W.S.CASSELS
63. Continuous semigroups in Banach algebras, A.M.SINCLAIR
64. Basic concepts of enriched category theory, G.M.KELLY
65. Several complex variables and complex manifolds I, M.J.FIELD
66. Several complex variables and complex manifolds II, M.J.FIELD
67. Classification problems in ergodic theory, W.PARRY & S.TUNCEL
68. Complex algebraic surfaces, A.BEAUVILLE
69. Representation theory, I.M.GELFAND et al.
70. Stochastic differential equations on manifolds, K.D.ELWORTHY
71. Groups - St Andrews 1981, C.M.CAMPBELL & E.F.ROBERTSON (eds.)
72. Commutative algebra: Durham 1981, R.Y.SHARP (ed.)
73. Riemann surfaces: a view towards several complex variables, A.T.HUCKLEBERRY
74. Symmetric designs: an algebraic approach, E.S.LANDER
75. New geometric splittings of classical knots (algebraic knots), L.SIEBENMANN & F.BONAHON
76. Linear differential operators, H.O.CORDES
77. Isolated singular points on complete intersections, E.J.N.LOOIJENGA
78. A primer on Riemann surfaces, A.F.BEARDON
79. Probability, statistics and analysis, J.F.C.KINGMAN & G.E.H.REUTER (eds.)
80. Introduction to the representation theory of compact and locally compact groups, A.ROBERT
81. Skew fields, P.K.DRAXL
82. Surveys in combinatorics: Invited papers for the ninth British Combinatorial Conference 1983, E.K.LLOYD (ed.)
83. Homogeneous structures on Riemannian manifolds, F.TRICERRI & L.VANHECKE
84. Finite group algebras and their modules, P.LANDROCK
85. Solitons, P.G.DRAZIN
86. Topological topics, I.M.JAMES (ed.)
87. Surveys in set theory, A.R.D.MATHIAS (ed.)
88. FPF ring theory, C.FAITH & S.PAGE
89. An F-space sampler, N.J.KALTON, N.T.PECK & J.W.ROBERTS
90. Polytopes and symmetry, S.A.ROBERTSON

London Mathematical Society Lecture Note Series: 77

Isolated Singular Points on Complete Intersections

E.J.N. LOOIJENGA
Professor of Mathematics, University of Nijmegen
The Netherlands

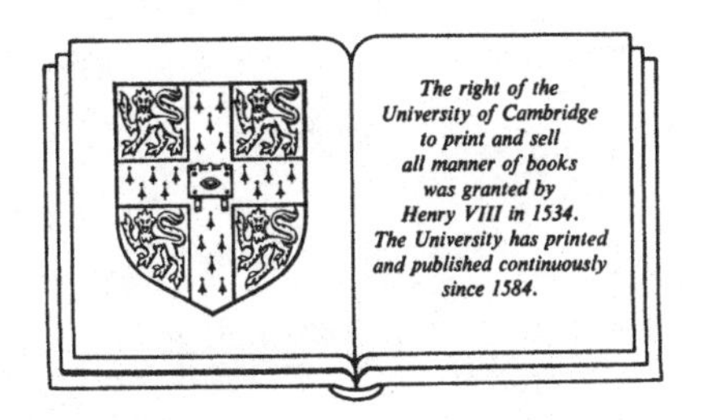
The right of the
University of Cambridge
to print and sell
all manner of books
was granted by
Henry VIII in 1534.
The University has printed
and published continuously
since 1584.

CAMBRIDGE UNIVERSITY PRESS
Cambridge
London New York New Rochelle
Melbourne Sydney

CAMBRIDGE UNIVERSITY PRESS
Cambridge, New York, Melbourne, Madrid, Cape Town, Singapore, São Paulo

Cambridge University Press
The Edinburgh Building, Cambridge CB2 8RU, UK

Published in the United States of America by Cambridge University Press, New York

www.cambridge.org
Information on this title: www.cambridge.org/9780521286749

First published 1984
Re-issued in this digitally printed version 2008

A catalogue record for this publication is available from the British Library

Library of Congress Catalogue Card Number: 82–9707

ISBN 978-0-521-28674-9 paperback

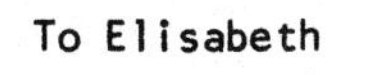

To Elisabeth

TABLE OF CONTENTS

INTRODUCTION

In the spring term of 1980 I gave a course on singularities at Yale University (while supported by NSF grant MCS 7905018), which provided the basis of a set of notes prepared for the first two years of the Singularity Intercity Seminar (1980 - 1982, at Leiden, Nijmegen and Utrecht, jointly run with Dirk Siersma and Joseph Steenbrink). These notes developed into the present book. As a consequence, aim and prerequisites of the seminar and this book are almost identical.

The purpose of the seminar was to introduce its participants to isolated singularities of complex spaces with particular emphasis on complete intersection singularities. When we started we felt that no suitable account was available on which our seminar could be based, so it was decided that I should supply notes, to be used by both the lecturers (in preparing their talks) and the audience. This was quite a purifying process: many errors and inaccuracies of the first draft were thus detected (and often corrected).

The prerequisites consisted of some algebraic and analytic geometry (roughly covering the contents of the books of Mumford (1976) and Narasimhan (1966)), some algebraic topology (as in Spanier (1966) and Godement (1958)) and some facts concerning Stein spaces. Given this background, my goal was to prove every assertion in the text. This has been achieved except for the coherence theorem (8.7) and some assertions in the descriptive chapter 1. An exception should also be made for the paragraphs marked with an asterisk (*): they generally give useful information, which however is not indispensable for what follows and so may be skipped. Perhaps the whole first chapter could have been marked with an asterisk. It gives interesting examples of isolated singularities (or of constructions thereof) with the purpose to indicate the position of complete intersection singularities among them and to describe material to which the theory is going to apply. It is mainly for the latter reason that this chapter should not be entirely skipped.

As each chapter has its own introduction. I shall not review the chapters separately, nor the whole book. I believe that the first seven chapters (with the exception of § 5.C) can be used as a basis for a course on the subject, assuming the audience has approximately the background described above. The contents of § 5.C and the last two chapters are somewhat more advanced and in addition, chapter 9 is of a more specialized nature. Some results or proofs may be new (at least do not appear in this form in the literature), examples are Ch. 1, p. 18ff, (4.7), (4.11), § 5.C, (6.13), (7.14), § 9.A, § 9.C. The references at the end should be regarded as a list of sources I consulted and not as a bibliography which aims to be complete in any respect.

Acknowledgements. Although all sources I used are cited, I want to single out some papers which were particularly useful to me: Lamotke (1975) for Ch. 3, Teissier (1976) for Ch. 4, Lê (1973, 1978) for Ch. 5 and Greuel (1975, 1980) for Ch. 8 and 9. As already mentioned, the book benefitted from criticism of the lecturers in our seminar. I mention in particular C. Cox, W. Janssen, P. Lemmens, P. Lorist, F. Menting,

G. Pellikaan, J. Stevens, D. van Straten and E. van Wijngaarden. Also, comments from the co-organizers, D. Siersma and J. Steenbrink, were very helpful. I take the occasion to thank the Dutch Organization for the Advancement of Pure Research (ZWO) for sponsoring our seminar and for supporting three of its participants. I am greatly indebted to W. Janssen for careful proofreading - his accurate job eliminated many errors and obscurities - and help in the exposition. I thank Ms. Ellen van Eldik for producing a beautiful camera ready typescript. Finally, I express my thanks to my wife, Elisabeth, for her continuous support during the writing of this book.

August 1983 E. Looijenga

1 EXAMPLES OF ISOLATED SINGULAR POINTS

Loosely speaking, an analytic germ (X,x) in $\mathbb{C}^N$ is called a complete intersection if the minimal number of equations by which it can be defined equals its codimension in $\mathbb{C}^N$. Although such germs will be our principal object of study, we must realize that quite often analytic germs are not given to us as the common zero set of a specific set of equations. In such cases it is unreasonable to expect these germs to be complete intersections. We illustrate this by describing several constructions of singular germs, most of which fail to yield complete intersections in general. Some of the germs which happen to be complete intersections, will reappear when we make a beginning of the classification in chapter 7. Another goal of this chapter is to make the reader acquainted with several interesting examples to which the theory we are going to develop may be applied. We do not always provide full proofs of the properties attributed to these singularities. The reader shouldn't feel uneasy about this, for in such cases we will not make any use of them.

1.A *Hypersurface singularities*

(1.1) Let X be an analytic set in an open $U \subset \mathbb{C}^{n+1}$ and let $x \in X$. The ideal $I_{X,x}$ of holomorphic functions at x vanishing on X is principal and nonzero if and only if each irreducible component of the germ (X,x) is of dimension n (see for instance Whitney (1972), Ch. 2, Thm's 10 C,D). We then say that (X,x) is a *hypersurface germ*. If $f \in m_{\mathbb{C}^{n+1},x}$ generates $I_{X,x}$, then the fact that I_X is a coherent O_U-module implies that there is an open neighbourhood U' of x in U such that f converges on U' and $I_X|U' = fO_{U'}$.

(1.2) *Proposition*. In this situation, the following are equivalent:

(i) There is a neighbourhood U'' of x in U such that $(X \cap U'')-\{x\}$ is nonsingular.

(ii) $(f,\frac{\partial f}{\partial z_0},\ldots,\frac{\partial f}{\partial z_n})O_{U,x} \supset m^k_{U,x}$ for some k.

(iii) $(\frac{\partial f}{\partial z_0},\ldots,\frac{\partial f}{\partial z_n})O_{U,x} \supset m^k_{U,x}$ for some k.

(iv) $\dim_{\mathbb{C}} O_{U,x}/(\frac{\partial f}{\partial z_0},\ldots,\frac{\partial f}{\partial z_n})O_{U,x} < \infty$.

(v) $\dim_{\mathbb{C}} O_{U,x}/(f,\frac{\partial f}{\partial z_0},\ldots,\frac{\partial f}{\partial z_n})O_{U,x} < \infty$.

Proof. (i) ⇒ (ii). If y is a nonsingular point of $X \cap U'$, then $\frac{\partial f}{\partial z_\nu}(y) \neq 0$ for some ν, for f generates $I_{X,y}$. So the common zero set of $f,\frac{\partial f}{\partial z_0},\ldots,\frac{\partial f}{\partial z_n}$ is contained in the singular locus $C_{X\cap U'}$ of $X \cap U'$. Since $C_{X\cap U''} \subset \{x\}$ (by assumption), the radical of $(f,\frac{\partial f}{\partial z_0},\ldots,\frac{\partial f}{\partial z_n})O_{U,x}$ is either $O_{U,x}$ or $m_{U,x}$ by the local analytic Nullstellensatz. This clearly implies (ii).

(ii) ⇒ (iii). Let $Y \subset U'$ denote the common zero set of $\frac{\partial f}{\partial z_0},\ldots,\frac{\partial f}{\partial z_n}$. It will be enough to show that $(Y,x) \subset (X,x)$ because the Nullstellensatz will then imply that $(\frac{\partial f}{\partial z_0},\ldots,\frac{\partial f}{\partial z_n})O_{U,x}$ and $(f,\frac{\partial f}{\partial z_0},\ldots,\frac{\partial f}{\partial z_n})O_{U,x}$ have the same radical. If $(Y,x) \not\subset (X,x)$, then we can find a germ of an analytic curve $s : (\mathbb{C},0) \to (Y,x)$ with $s(t) \notin X$ for $t \neq 0$. But

$$\frac{d}{dt}(f\circ s) = \Sigma^n_{\nu=0}(\frac{\partial f}{\partial z_\nu}\circ s)\frac{ds_\nu}{dt} = 0$$

and so $f \circ s$ is constant equal to $f \circ s(0) = 0$. This contradicts our assumption that $s(t) \notin X$ for $t \neq 0$.

(iii) ⇒ (iv), (iv) ⇒ (v) and (ii) ⇒ (i) are easy, while (v) ⇒ (ii) will follow from the lemma below applied to $M = \mathcal{O}_{U,x}/(f, \frac{\partial f}{\partial z_0}, \ldots, \frac{\partial f}{\partial z_n})\mathcal{O}_{U,x}$.

(1.3) *Lemma.* If M is an $\mathcal{O}_{U,x}$-module of finite $\mathbb{C}$-dimension d, then $m^d_{U,x}$ annihilates M.

Proof. Put $d_k := \dim_{\mathbb{C}} M/m^k_{U,x}M$, $k = 0,1,2,\ldots$. Then $0 = d_0 \leq d_1 \leq \ldots$ $\ldots \leq d_k \leq \ldots \leq d$. So for some $k \leq d$, we have $d_k = d_{k+1}$. This means that $m^k_{U,x}M = m^{k+1}_{U,x}M$. Since $m^k_{U,x}M \subset M$ is of finite $\mathbb{C}$-dimension, $m^k_{U,x}M$ is a noetherian $\mathcal{O}_{U,x}$-module so that Nakayama's lemma applies: it follows that $m^k_{U,x}M = 0$.

(1.4) If one of the (equivalent) conditions of (1.2) is satisfied we say that (X,x) is an *isolated hypersurface singularity*. The dimension occurring in (1.2)-iv is usually called the *Milnor number* of X at x and denoted $\mu(X,x)$. Milnor (1968) originally defined this number in a topological manner but we will see in chapter 5 that the two definitions agree. The dimension in (1.2)-v will be interpreted in chapter 6 when we investigate the deformation theory of (X,x). We follow Greuel and call it the *Tjurina number* of (X,x), denoted $\tau(X,x)$. Clearly $\tau(X,x) \leq \mu(X,x)$ and we have equality if and only if $f \in (\frac{\partial f}{\partial z_0}, \ldots, \frac{\partial f}{\partial z_n})\mathcal{O}_{U,x}$. This is for instance the case if f is *weighted homogeneous*. This means that there exist positive integers $d_0,\ldots,d_n,N$ such that if we give z_ν degree d_ν, f is a homogeneous polynomial of degree N, in other words f is of the form

$$f(z) = \Sigma_{i_0 d_0 + \ldots + i_n d_n = N} \; a_{i_0 \ldots i_n} z_0^{i_0} \ldots z_n^{i_n}.$$

(We take $x = 0$.) Then it is easily checked that

$$f(z) = \Sigma_{\nu=0}^{n} \frac{d_\nu}{N} z_\nu \frac{\partial f}{\partial z_\nu} .$$

The condition that $f \in (\frac{\partial f}{\partial z_0}, \ldots, \frac{\partial f}{\partial z_n})\mathcal{O}_{U,x}$ is coordinate invariant and hence also satisfied for an f which is weighted homogeneous with respect to some coordinate system at $(\mathbf{C}^{n+1},x)$. According to Saito (1971) there is a converse to this: if f defines an isolated hypersurface singularity and $f \in (\frac{\partial f}{\partial z_0}, \ldots, \frac{\partial f}{\partial z_n})\mathcal{O}_{U,x}$, then f is weighted homogeneous with respect to some coordinate system.

1.B *Complete intersections*

(1.5) Let x be a point of an analytic set X, defined in an open $U \subset \mathbf{C}^N$, and let n denote the dimension of X at x. Then near x, X cannot be defined as the common zero set of fewer than N-n holomorphic functions at x. If we can do it with N-n such functions, then we say that X is a *geometric complete intersection* at x. This is a nontrivial condition: for instance the union of the (z_1,z_2)-plane and the (z_3,z_4)-plane in $\mathbf{C}^4$ is not a geometric complete intersection at the origin, see for instance Gunning (1974), p. 159. Likewise if $I \subset m_{\mathbf{C}^N,x}$ is an ideal which defines a germ of dim n, then we say that I defines a *complete intersection* at x if I admits N-n generators $f_1,\ldots,f_{N-n}$. The following result characterizes such (N-n)-tuples algebraically.

(1.6) Let $f_1,\ldots,f_{N-n} \in m_{\mathbf{C}^N,x}$ generate an ideal I in $\mathcal{O}_{\mathbf{C}^N,x}$. Then I defines a complete intersection of dim n if and only if $f_1,\ldots,f_{N-n}$ is an $\mathcal{O}_{\mathbf{C}^N,x}$*-sequence*, i.e. f_j is not a zero-divisor of $\mathcal{O}_{\mathbf{C}^N,x}/(f_1,\ldots,f_{j-1})\mathcal{O}_{\mathbf{C}^N,x}$ for $j = 1,\ldots,N-n$. If either condition is fulfilled, $\mathcal{O}_{\mathbf{C}^N,x}/I$ is a Cohen-

Macaulay ring of dim n. In particular, $\dim \mathcal{O}_{\mathbb{C}^N,x}/P = n$ for any associated prime ideal P of $\mathcal{O}_{\mathbb{C}^N,x}/I$.

For a proof, see for instance Matsumura (1980), Th. 30.

The property that an ideal $I \subset m_{\mathbb{C}^N,x}$ defines a complete intersection at x only depends on the $\mathbb{C}$-algebra $\mathcal{O}_{\mathbb{C}^N,x}/I$, as will follow from (1.8) below. Let us start with proving an intermediate result, which we shall use at other places as well.

(1.7) *Lemma*. Let $I \subset \mathcal{O}_{\mathbb{C}^N,0}$ and $J \subset \mathcal{O}_{\mathbb{C}^N,0}$ be ideals and assume we are given an isomorphism of $\mathbb{C}$-algebras $\phi^* : \mathcal{O}_{\mathbb{C}^N,0}/J \to \mathcal{O}_{\mathbb{C}^N,0}/I$. Then there exists an analytic automorphism $\Phi : (\mathbb{C}^N,0)\circlearrowleft$ with $\Phi^*(J) = I$ such that Φ^* induces ϕ^*.

Proof. We assume that I and J are both distinct from $\mathcal{O}_{\mathbb{C}^N,0}$. Let r denote the dimension of the $\mathbb{C}$-vector space $(I+m^2)/m^2$ (where $m = m_{\mathbb{C}^N,0}$). Choose $z'_1,\ldots,z'_r \in I$ such that their images in $(I+m^2)/m^2$ give a basis and extend this set to a coordinate system $z'_1,\ldots,z'_N$ for $(\mathbb{C}^N,0)$. In the exact sequence

$$0 \to (I+m^2)/m^2 \to m/m^2 \to m/(I+m^2) \to 0$$

the term $m/(I+m^2)$ is naturally isomorphic to m_I/m_I^2 where $m_I \subset \mathcal{O}_{\mathbb{C}^N,0}/I$ denotes the maximal ideal of $\mathcal{O}_{\mathbb{C}^N,0}/I$. Since $\mathcal{O}_{\mathbb{C}^N,0}/I$ and $\mathcal{O}_{\mathbb{C}^N,0}/J$ are isomorphic, it follows that $m/(I+m^2)$ and $m/(J+m^2)$ have the same $\mathbb{C}$-dimension. Hence $\dim_{\mathbb{C}}(J+m^2)/m^2 = r$. Repeating the above construction with J yields a coordinate system $z''_1,\ldots,z''_N$ for $(\mathbb{C}^N,0)$ with $z''_1,\ldots,z''_r \in J$. Choose $\Phi_\nu \in m_{\mathbb{C}^N,0}$ $(\nu = r+1,\ldots,N)$ such that its reduction mod I is just the image of $z''_\nu + J$ under ϕ^*. Define $\Phi : (\mathbb{C}^N,0)\circlearrowleft$ by $\Phi^*(z''_\nu) = z'_\nu$ if $\nu \leq r$ and $\Phi^*(z''_\nu) = \Phi_\nu$ if $\nu > r$. Then Φ induces a map of exact sequences

$$\begin{array}{ccccccccc} 0 \to & (J+m^2)/m^2 & \to & m/m^2 & \to & m_J/(m_J)^2 & \to 0 \\ & \downarrow & & \downarrow & & \downarrow & \\ 0 \to & (I+m^2)/m^2 & \to & m/m^2 & \to & m_I/(m_I)^2 & \to 0 \end{array}$$

The vertical map on the left is an isomorphism by construction, whereas the one on the right is so because it is induced by ϕ^*. Hence the middle map is an isomorphism. As this represents the derivative of Φ in 0 it follows that Φ is an isomorphism. Clearly, $\Phi^*(J) \subset I$. Since Φ^* induces ϕ^*, we have in fact $\Phi^*(J) = I$.

(1.8) *Lemma*. Let $I \subset m_{\mathbb{C}^N,0}$ and $J \subset m_{\mathbb{C}^M,0}$ be ideals such that $A := \mathcal{O}_{\mathbb{C}^N,0}/I$ and $B := \mathcal{O}_{\mathbb{C}^M,0}/J$ are isomorphic $\mathbb{C}$-algebras. If $d(I)$ respectively $d(J)$ denotes the minimal number of generators of I respectively J, then $N-d(I) = M-d(J)$.

Proof. Put $r := \dim_{\mathbb{C}}(I+m^2)/m^2$. If I admits the generators $g_1,\ldots,g_{d(I)} \in m_{\mathbb{C}^N,0}$ then by making a suitable coordinate transformation of $(\mathbb{C}^N,0)$, we may assume that $g_i = z_{N-r+i}$ for $i \leq r$ and $g_i \in m^2_{\mathbb{C}^N,0}$ for $i > r$. Let $\pi : \mathcal{O}_{\mathbb{C}^N,0} \to \mathcal{O}_{\mathbb{C}^{N-r},0}$ denote the obvious projection and set $I' = \pi(I)$. Then $\pi^{-1}(I') = I$, so that π induces an isomorphism of $\mathcal{O}_{\mathbb{C}^N,0}/I$ onto $\mathcal{O}_{\mathbb{C}^{N-r},0}/I'$. As I' is generated by $\pi(g_{r+1}),\ldots,\pi(g_{d(I)})$, it follows that $d(I') \leq d(I)-r$. On the other hand, if $g',\ldots,g'_{d(I')}$ generate I', then lifts of these in I toghether with $z_{N-r+1},\ldots,z_N$ generate I, so that $d(I) \leq d(I')+r$, also. It follows that $d(I') = d(I)-r$, where I' has now the virtue of being contained in $m^2_{\mathbb{C}^{N-r},0}$. So without loss of generality we may assume that I and J are in the squares of the maximal ideals. We claim that then $N = M$. This simply follows from the fact that $m_{\mathbb{C}^N,0}/m^2_{\mathbb{C}^N,0}$ maps isomorphically onto $m_{\mathbb{C}^N,0}/(I+m^2_{\mathbb{C}^N,0}) \cong m_A/m_A^2$, where m_A denotes the maximal ideal of A, so that N is invariantly characterized as $\dim_{\mathbb{C}} m_A/m_A^2$. Similarly, $M = \dim_{\mathbb{C}} m_B/m_B^2$ and so $N = M$. The lemma now follows from (1.7).

(1.9) The preceding argument shows that for any local analytic algebra A (i.e. a $\mathbb{C}$-algebra isomorphic to one of the form $\mathcal{O}_{\mathbb{C}^N,0}/I$) the minimal N such that A is isomorphic to some quotient of $\mathcal{O}_{\mathbb{C}^N,0}$ is given by $\dim_{\mathbb{C}} m_A/m_A^2$. This number is called the *embedding dimension* of A. The difference $\dim m_A/m_A^2 - \dim A$ is called the *embedding codimension* of A. We say that a local analytic $\mathbb{C}$-algebra A is a *(local) complete intersection algebra* if for some surjection $\pi : \mathcal{O}_{\mathbb{C}^N,x} \to A$ of local $\mathbb{C}$-algebras, $I := \mathrm{Ker}(\pi)$ defines a complete intersection (X,x) in the previous sense (by the preceding lemma, this is then so for *any* such π). The case that concerns us most is when X has an isolated singular point or is regular in x. This means that if $f_1,\dots,f_{N-n}$ is a set of generators of I (with $n = \dim A$) then there is an open neighbourhood V of x in $\mathbb{C}^N$ on which $f_1,\dots,f_{N-n}$ converge and for all $y \neq x$ in the common zero set X of $f_1,\dots,f_{N-n}$, $df_1(y),\dots,df_{N-n}(y)$ are linearly independent. This is also equivalent to the condition that the ideal in $\mathcal{O}_{\mathbb{C}^N,x}$ generated by $f_1,\dots,f_{N-n}$ and the determinants of the $(N-n)\times(N-n)$ submatrices of $(\frac{\partial f_j}{\partial z_\nu})$ contains a power of $m_{\mathbb{C}^N,x}$. We then say that (X,x) endowed with its local $\mathbb{C}$-algebra $\mathcal{O}_{\mathbb{C}^N,x}/I$ ($\cong$ A) is an *isolated complete intersection singularity* (so this includes the case that $\mathcal{O}_{\mathbb{C}^N,x}/I$ is regular). Henceforth we shall abbreviate this as *icis*. We will often use this in expressions like:
$(f_1,\dots,f_{N-n}) : (\mathbb{C}^N,x) \to (\mathbb{C}^{N-n},0)$ (or $I \subset \mathcal{O}_{\mathbb{C}^N,x}$, or A) defines an icis.

(1.10) *Proposition.* If $I \subset \mathcal{O}_{\mathbb{C}^N,x}$ defines an icis of dim $n > 0$, then I is its own radical.

Proof. Let $f_1,\dots,f_{N-n} \in I$, V and X be as above. We want to show that the sheaf $F := I_X/(f_1,\dots,f_{N-n})\mathcal{O}_V$ is trivial. This is clearly the case outside {x}. Since F is a coherent sheaf of $\mathcal{O}_V$-modules, its annihilator $\mathrm{Ann}(F) = \{f \in \mathcal{O}_V : fF = 0\}$ is also coherent. Since $\mathrm{Ann}(F_y) = \mathcal{O}_{V,y}$ for

$y \neq x$, we must have $\mathrm{Ann}(F_x) \supset m_{V,x}^{\ell}$ for some ℓ (by the Nullstellensatz). Either $\ell = 0$ (and hence $F_x = \{0\}$ as was to be shown) or $\ell > 0$ and then each element of F_x must be a zero-divisor. According to (1.6) this can only happen when $n = 0$.

Example 1. The pair of quadratic forms in $\mathbf{C}^N$ $(N \geq 2)$,

$$f_1(z_1,\ldots,z_N) = z_1^2+\ldots+z_N^2$$

$$f_2(z_1,\ldots,z_N) = \lambda_1 z_1^2+\ldots+\lambda_N z_N^2$$

defines an icis of dim N-2 at $0 \in \mathbf{C}^N$ if and only if the coefficients $\lambda_1,\ldots,\lambda_N$ are all distinct. This illustrates a result due to Hamm (1969) which says that for a given $(N-n)\times N$ matrix $(n_{j\nu})$ with $n_{j\nu} \in \mathbf{N}$, the equations

$$f_j(z_1,\ldots,z_N) = \Sigma_{\nu=1}^N a_{j\nu} z_\nu^{n_{j\nu}}$$

define an icis for almost any coefficient matrix $(a_{j\nu})$.

1.C *Quotient singularities*

Let G be a finite group of local analytic automorphisms of $\mathbf{C}^n$ at 0. Following Cartan (1957), this action can be linearized, that is, in terms of a (possibly) new coordinate system for $(\mathbf{C}^n,0)$, G will act linearly. Therefore it is no restriction to assume that G is a subgroup of $GL_n(\mathbf{C})$. Then G also acts on $\mathbf{C}[z_1,\ldots,z_n]$ by $(g.\phi)(z) = \phi(g^{-1}(z))$. The G-invariant polynomials form a homogeneous subalgebra $\mathbf{C}[z_1,\ldots,z_n]^G$ of $\mathbf{C}[z_1,\ldots,z_n]$, which is finitely generated and normal, see (Bourbaki: AC V), §1, no. 9. Choose homogeneous generators $\phi_1,\ldots,\phi_N$ of $\mathbf{C}[z_1,\ldots,z_n]^G$ of positive degree and define a polynomial mapping $\Phi : \mathbb{C}^n \to \mathbb{C}^N$ by $\Phi(z) = (\phi_1(z),\ldots,\phi_N(z))$. Then Φ is constant on the G-orbits and thus factors through the orbit space $G\backslash\mathbb{C}^n$ by a mapping $\Phi' : G\backslash\mathbb{C}^n \to \mathbb{C}^N$.

(1.11) *Proposition.* The mapping Φ' is a proper homeomorphism of $G\backslash\mathbb{C}^n$ onto a normal algebraic subvariety of $\mathbb{C}^N$ whose algebra of regular functions corresponds under Φ to $\mathbb{C}[z_1,\ldots,z_n]^G$.

Proof. Let us first show that Φ' is injective. For this it suffices to check that given $z,z' \in \mathbb{C}^n$, $z' \notin G.z$, we have $\Phi(z) \neq \Phi(z')$. Choose a polynomial $\psi \in \mathbb{C}[z_1,\ldots,z_n]$ with $\psi(z') = 0$ and $\psi(g.z) \neq 0$ for all $g \in G$ and put $\psi_0 := \pi_{g\in G}(g.\psi)$. Then $\psi_0 \in \mathbb{C}[z_1,\ldots,z_n]^G = \mathbb{C}[\phi_1,\ldots,\phi_N]$. Since $\psi_0(z') = 0 \neq \psi_0(z)$, we must have $\phi_\nu(z) \neq \phi_\nu(z')$ for some ν.

Next we show that Φ is proper. Since Φ' is injective, we have $\Phi'^{-1}(0) = G.\{0\} = \{0\}$. Since the unit sphere of $\mathbb{C}^n$ is compact there exists an $\varepsilon > 0$ such that $\max\{|\phi_1(z)|,\ldots,|\phi_\nu(z)|\} \geq \varepsilon$ if $|z| = 1$. If d_ν is the degree of ϕ_ν, then it follows that

$$|\Phi(z)|^2 = |z|^{2d_1}|\phi_1(\tfrac{z}{|z|})|^2 + \ldots + |z|^{2d_N}|\phi_N(\tfrac{z}{|z|})|^2$$

$$\geq \varepsilon^2 \min.\{|z|^{2d_1},\ldots,|z|^{2d_N}\}.$$

This shows that the pre-image of a bounded set under Φ is bounded and hence that Φ is proper.

Since Φ' is a proper continuous injection, it is a homeomorphism onto its image. A basic result in elimination theory, e.g. Mumford (1976), (2.31), (2.33), implies that the image X of the proper algebraic mapping $\Phi = \mathbb{C}^n \to \mathbb{C}^N$ is algebraic in $\mathbb{C}^N$. Clearly, Φ^* maps the algebra of regular functions on X isomorphically onto $\mathbb{C}[z_1,\ldots,z_n]^G$. The latter is normal and hence so is X.

This result can be understood in a more conceptual way as saying that the orbit space $G\backslash\mathbb{C}^n$ is in a natural manner a normal affine algebraic variety whose algebra of regular functions is $\mathbb{C}[z_1,\ldots,z_n]^G$. From now on, we shall view $G\backslash\mathbb{C}^n$ as such. The germ of $G\backslash\mathbb{C}^n$ at G.0 (and any analytic germ isomorphic to such a germ) is called a *quotient singularity*. It is clear that $G\backslash\mathbb{C}^n$ is isomorphic to $\mathbb{C}^n$ if and only if $\mathbb{C}[z_1,\ldots,z_n]^G$ is a polynomial algebra. There is a beautiful characterization of such G due to Chevalley (1955): $\mathbb{C}[z_1,\ldots,z_n]^G$ is a polynomial algebra if and only if G is generated by complex reflections ($g \in GL_n(\mathbb{C})$ is called a *complex reflection* if it is of finite order $\neq 1$ and leaves a hyperplane pointwise fixed). (This reduces the study of quotient singularities $G\backslash\mathbb{C}^n$ to the case where G contains no complex reflections. For if $H \subset G$ denotes the subgroup generated by all complex reflections in G, then H is normal in G and the factor group G/H will act on $H\backslash\mathbb{C}^n$ - this corresponds to the natural action of G/H on $\mathbb{C}[z_1,\ldots,z_n]^H$ - and $G\backslash\mathbb{C}^n$ is identified with the orbit space $(G/H)\backslash\mathbb{C}^n$. It can be shown that G/H contains no complex reflection.) We consider the case when $G \subset SL_2(\mathbb{C})$ in more detail. Notice that then G doesn't contain complex reflections: if $g \in SL_2(\mathbb{C})$ leaves a line pointwise fixed, then it is unipotent and hence of infinite order unless $g = 1$. Since $G\backslash\mathbb{C}^2$ is normal, it will have G.0 as an isolated singular point if $G \neq \{1\}$.

Example 2. Let m be a positive integer and let $G = \{\begin{pmatrix}\zeta & 0\\ 0 & \zeta^{-1}\end{pmatrix} : \zeta^m = 1\}$. Then $\mathbb{C}[z_1,z_2]^G$ is generated by z_1^m, z_2^m, z_1z_2. We use these as the coordinates of $f : \mathbb{C}^2 \to \mathbb{C}^3$. Clearly, the image of f is contained in the hypersurface defined by $t_1t_2-t_3^m = 0$ As this hypersurface is irreducible and $\dim f(\mathbb{C}^2) = 2$, the two must coincide. So the germ $(G\backslash\mathbb{C}^2, G.0)$ is isomorphic to the hypersurface singularity defined by $t_1t_2-t_3^m$. It is customary to denote the isomorphism type of this germ by A_{m-1}. An A_1-singularity is also called a *quadratic singularity.*

Example 3. (Klein, 1884). Let $2 \leq p \leq q \leq r$ be integers such that $\frac{1}{p}+\frac{1}{q}+\frac{1}{r} > 1$ (so $(p,q,r) = (2,2,r),(2,3,3),(2,3,4)$ or $(2,3,5)$). Then according to spherical geometry there exists a solid spherical triangle Δ on the unit sphere $S^2 \subset \mathbf{R}^3$ whose angles are $\frac{\pi}{p}, \frac{\pi}{q}, \frac{\pi}{r}$. The orthogonal reflections in the sides of Δ generate a finite subgroup Σ of the orthogonal group $O_3(\mathbf{R})$ which has Δ as a fundamental domain (so $G.\Delta = S^2$ and if $g \in G-\{1\}$, then $g(\mathring{\Delta}) \cap \mathring{\Delta} = \emptyset$). The truth of this statement is easily visualized with the help of the regular polyhedra (Δ is obtained by central projection of the shaded area).

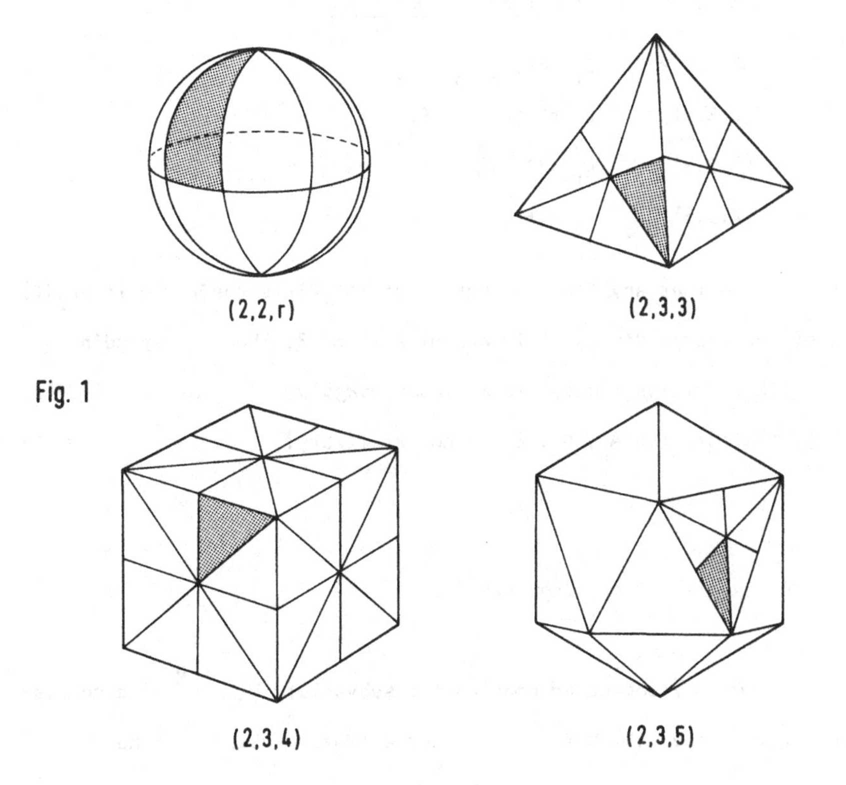

Fig. 1

The subgroup $\Sigma_+ := \Sigma \cap SO_3(\mathbb{R})$ of Σ (of index 2) is just the group of motions of the corresponding polyhedron. Now, SU_2 naturally acts on $\mathbb{C}^2$ and thereby on the complex projective line $\mathbb{P}^1 = \mathbb{C} \cup \{\infty\}$. Stereographic projection defines an isomorphism between $\mathbb{P}^1$ and the unit sphere S^2 of $\mathbb{R}^3$ which respects the conformal structure and this determines a homomorphism $\rho : SU_2 \to SO_3(\mathbb{R})$ of Lie groups. It is well known (and easy to check) that $\mathrm{Ker}(\rho) = \{\pm 1\}$. Since SU_2 and $SO_3(\mathbb{R})$ are compact connected and have the same dimension (3), it follows that ρ is a two-fold covering. We put $G := \rho^{-1}(\Sigma_+)$. It appears that $\mathbb{C}[z_1,z_2]^G$ is generated by three elements, see Klein (1884) and Milnor (1975). By the argument used in the cyclic case, it follows that $\mathbb{C}^2/G$ is isomorphic to a hypersurface in $\mathbb{C}^3$. These are

(p,q,r)	equation	notation
$(2,2,r \geq 2)$	$t_1^{r+1}+t_1t_2^2+t_3^2$	D_{r+2}
$(2,3,3)$	$t_1^4+t_2^3+t_3^2$	E_6
$(2,3,4)$	$t_1^3t_2+t_2^3+t_3^2$	E_7
$(2,3,5)$	$t_1^5+t_2^3+t_3^2$	E_8

It can be shown that any finite subgroup of $SL_2(\mathbb{C})$ is conjugate in $SL_2(\mathbb{C})$ to one of the groups discussed in examples 2 and 3. The corresponding quotient singularities are called *Kleinian singularities*. As we will find in Ch. 7, the notation A_μ, D_μ, E_μ is not arbitrary.

1.D *Quasi-conical singularities*

For any connected nonsingular subvariety V of $\mathbb{P}^N$, the corresponding affine cone $C(V) \subset \mathbb{C}^{N+1}$ is either a linear subspace or has

$0 \in C(V)$ as its unique singular point. This simple way of getting isolated singular points can be generalized as follows. Let V be a nonsingular projective variety endowed with an ample line bundle ℓ over V. (This means that ℓ^k is very ample for some $k > 0$, i.e. there exist sections $\sigma_0,\ldots,\sigma_N$ of ℓ^k such that $[\sigma_0,\ldots,\sigma_N] : V \to \mathbf{P}^N$ is everywhere defined and embeds V in $\mathbf{P}^N$.) Notice that a section σ of ℓ^k may be viewed as a function on the total spece $\mathrm{Tot}(\ell^{-1})$ of the dual of ℓ satisfying the homogeneity condition $\sigma(\lambda.z) = \lambda^k\sigma(z)$, $(\lambda \in \mathbf{C},\ z \in \mathrm{Tot}(\ell^{-1}))$. Since such a function is constantly zero on the zero section of ℓ^{-1} (when $k > 0$), we may think of the graded algebra $A^* := \oplus_{k=0}^{\infty} \Gamma(\ell^k)$ as an algebra of functions on the pointed space $(C(\ell^{-1}),*)$ obtained from $\mathrm{Tot}(\ell^{-1})$ by collapsing its zero section to a point. The ampleness of ℓ implies that A separates the points of $C(\ell^{-1})$. Furthermore, it can be shown that A is a normal $\mathbf{C}$-algebra of finite type. So $C(\ell^{-1})$ has the structure of an affine algebraic variety which is smooth outside the vertex.

Example 4. Let ℓ be the standard line bundle over $\mathbf{P}^1$ of (positive) degree d (its global sections are the homogeneous polynomials of degree d in $\mathbf{C}[z_0,z_1]$). Then ℓ is ample and $C(\ell^{-1})$ can be identified with the image of the map $(z_0,z_1) \in \mathbf{C}^2 \to (z_0^d,z_0^{d-1}z_1,\ldots,z_1^d) \in \mathbf{C}^{d+1}$. For $d = 2$ this gives us back the quadratic singularity A_1, but for $d > 2$ we do not get a complete intersection.

Example 5. Let E be an elliptic curve and let ℓ be a line bundle over E of degree $d > 0$. Then ℓ is ample. The singularity of $C(\ell^{-1})$ at its vertex (or rather its isomorphism type) is called a *simply-elliptic singularity* of *degree* d. Following K. Saito (1974) we get a hypersurface if $d = 1,2,3$, for $d = 4$ a complete intersection in $\mathbf{C}^4$, but for $d > 4$, $(C(\ell^{-1}),*)$ is not an icis. We give equations for $d \leq 4$ (op. cit.)

degree	equation(s)	notation
1	$z_1^2+z_2^3+z_3^6+az_1z_2z_3$	$\tilde{E}_8$ or $T_{2,3,6}(a)$
2	$z_1^2+z_2^4+z_3^4+az_1z_2z_3$	$\tilde{E}_7$ or $T_{2,4,4}(a)$
3	$z_1^3+z_2^3+z_3^3+az_1z_2z_3$	$\tilde{E}_6$ or $T_{3,3,3}(a)$
4	$(z_1^2+z_2^2+az_3z_4, z_1z_2+z_3^2+z_4^2)$	$\tilde{D}_5$ or $T_{2,2,2,2}(a)$

In all cases the j-invariant of E is a rational function of a (so a should avoid the values for which this function is not defined). The notation $\tilde{E}_8,\ldots,\tilde{D}_5$ will be explained in Ch. 7.

We may combine the constructions of (1.C) and (1.D) by starting out from a connected projective variety V, an ample line bundle ℓ over V and a finite group G of linear automorphisms of ℓ (by this we mean that G commutes with scalar multiplication). Then G acts on $C(\ell^{-1})$ and in much the same way as for quotient singularities it is shown that the orbit space $X := G\backslash C(\ell^{-1})$ is in a natural way a normal affine algebraic variety whose algebra of regular functions is the G-invariant part of $\oplus_{k=0}^{\infty}\Gamma(\ell^k)$. Notice that X inherits from $C(\ell^{-1})$ a $\mathbf{C}^*$-action with the property that $\lambda.x \to *$ as $|\lambda| \to \infty$. Any affine variety with $\mathbf{C}^*$-action which arises in this manner will be called a *quasi-cone*. (N.B. Our terminology differs from that of Dolgachev (1975): he reserves the notion quasi-cone for (what we will later call) singularities with good $\mathbf{C}^*$-action.)

Example 6. Let $p \leq q \leq r$ be positive integers such that $\frac{1}{p}+\frac{1}{q}+\frac{1}{r} < 1$. Then the Poincaré upper half plane $H = \{z = x+iy \in \mathbf{C} : y > 0\}$, endowed with the $PSL_2(\mathbf{R})$-invariant metric $y^{-2}(dx^2+dy^2)$, contains a solid triangle Δ with angles $\frac{\pi}{p}, \frac{\pi}{q}, \frac{\pi}{r}$. The group Σ of automorphisms of H generated by reflections in the edges of Δ acts properly discontinuously on H and has Δ as a fundamental domain, ((Milnor (1975)).Let $\Sigma_+ \subset \Sigma$ be the subgroup of index two which preserves the orientation. Then $\Sigma_+ \subset PSL_2(\mathbf{R})$ and Σ_+ acts linearly on the tangent bundle TH of H. Denote by X the space obtained by collapsing the image of the zero section in $\Sigma_+\backslash TH$ to a point. We prove that X is of the type described above. According to Fox (1952), Σ_+ contains a normal subgroup of finite index which has no elements of finite order. This implies that N acts freely on H. The orbit space $V := N\backslash H$ will be a nonsingular compact curve on which the translates of Δ induce a triangulation with $(\frac{1}{p} + \frac{1}{q} + \frac{1}{r}).|\Sigma_+/N|$ vertices, $3|\Sigma_+/N|$ edges and $2|\Sigma_+/N|$ triangles. It follows that the euler characteristic of V is $(\frac{1}{p} + \frac{1}{q} + \frac{1}{r} - 1)|\Sigma_+/N| < 0$ (this also follows from the fact that H is a universal cover of V so that V must have genus ≥ 2). As is well-known, the cotangent bundle τ_V^* of V is then ample, so that $C(\tau_V)$ exists as an affine algebraic variety. The finite group $G := \Sigma_+/N$ acts on $C(\tau_V)$ and it is clear that X can be identified with the G-orbit space of $C(\tau_V)$. The resulting isolated singularity is called a *triangle singularity* and denoted $D_{p,q,r}$. Following Dolgachev (1974) for exactly 14 triples (p,q,r), this is a hypersurface singularity. Pinkham (1977c) showed that for precisely 8 other triples we still have a complete intersection (in $\mathbf{C}^4$). An alternate description of these singularities will be given at the end of this chapter.

Example 7. Let X be a connected normal affine algebraic surface endowed with a $\mathbf{C}^*$-action and a distinguished point $x \in X$ such that for all $z \in X$, $\lambda.z$ tends to x as $|\lambda| \to \infty$. Then Dolgachev and Pinkham (1977a) prove that there exists a smooth compact curve V, an ample line bundle ℓ over V and a finite group G of linear automorphisms of ℓ such that X and $G\backslash C(\ell^*)$ are isomorphic as C^*-varieties.

1.E *Cusp singularities*

When compactifying certain orbit spaces, (isolated) singularities may arise in the following manner. Suppose we are given an open convex non-void cone C in R^n which doesn't contain a linear subspace $\neq \{0\}$ and a subgroup G of $SL_n(\mathbb{Z})$ which preserves C and acts properly on it. Assume moreover that there is a compact subset K of C such that any half line in C meets $g(K)$ for some $g \in G$ (in other words, the G-orbit space of the corresponding subset of $P(R^n)$ is compact). Let $\mathcal{D}$ denote the domain $\{\xi + iy \in \mathbb{C}^n/\mathbb{Z}^n : y \in C\}$. Then G acts (by complexification) on $\mathcal{D}$ and this action is proper. So $G\backslash\mathcal{D}$ is in a natural way a normal analytic space (Cartan (1957)). This space admits sufficiently many holomorphic functions to realize it as an analytic set: if $\phi : R^n \to R$ is an integral linear form which is positive on C, then the Fourier series

$$S_\phi := \Sigma_{\phi' \in G.\{\phi\}} e^{2\pi i \phi'}$$

converges absolutely and uniformly on compact subsets of $\{x + iy \in \mathbb{C}^n : y \in C\}$. Since S_ϕ is invariant under the translation group $\mathbb{Z}^n$ as well as under G it defines a holomorphic function S'_ϕ on $G\backslash\mathcal{D}$. A finite number N of these embed $G\backslash\mathcal{D}$ as an analytic subset of an open subset of $\mathbb{C}^N$. It turns out that we can take this open set to be a punctured neighbourhood of 0 in $\mathbb{C}^N$. More precisely, let ∞ be an abstract point and extend the topology of $G\backslash\mathcal{D}$ to the disjoint union $X := G\backslash\mathcal{D} \cup \{\infty\}$ by postulating that the sets $G\backslash\{\xi + iy \in \mathbb{C}^n/\mathbb{Z}^n : y \in U\} \cup \{\infty\}$, with U running over the open non-void convex G-invariant subsets of C, form a neighbourhood basis of ∞ in X. It is not hard to show that the functions S'_ϕ extend continuously in ∞ (with $S'_\phi(\infty) = 0$ if $\phi \neq 0$) and that X thus acquires the structure of a normal analytic space. The germ of X at ∞ (and any analytic germ isomorphic to it) is called a *cusp singularity*. If G acts freely on C, then $G\backslash\mathcal{D}$ is non-singular and so (X,∞) is either regular or an isolated singularity.

Example 8. For n = 1, the only example is $C = \{x \in \mathbf{R} : x > 0\}$ and $G = \{1\}$. Then $G\backslash\mathcal{D} = \mathcal{D} = H/\mathbf{Z}$. If we identify $H/\mathbf{Z}$ with the punctured complex unit disc $\Delta^* = \Delta - \{0\} \subset \mathbf{C}$ via the map $t \in H/\mathbf{Z} \to e^{2\pi i t} \in \Delta^*$ then the inclusion $G\backslash\mathcal{D} \subset G\backslash\mathcal{D} \cup \{\infty\}$ corresponds to $\Delta^* \subset \Delta$.

Example 9. Let $\sigma \in SL_2(\mathbf{Z})$ have two distinct positive eigenvalues (this is equivalent to trace $(\sigma) > 2$) and let $C \subset \mathbf{R}^2$ be an open sector bounded by two distinct σ-invariant half lines. Then $\sigma(C) = C$ and the (infinite cyclic) subgroup G of $SL_2(\mathbf{Z})$ generated by σ acts freely and properly on C. We leave it to the reader to verify that for $y \in C$ the segment $K = [y, \sigma(y)]$ has the property that the G-orbit of any half line in C meets K. Following Karras (1977), (X,∞) is a hypersurface germ if and only if $\sigma^{\pm 1}$ is conjugate in $SL_2(\mathbf{Z})$ to

$$\begin{pmatrix} 0 & 1 \\ -1 & p \end{pmatrix}\begin{pmatrix} 0 & 1 \\ -1 & q \end{pmatrix}\begin{pmatrix} 0 & 1 \\ -1 & r \end{pmatrix}$$

for certain positive integers p,q,r with $\frac{1}{p} + \frac{1}{q} + \frac{1}{r} < 1$ (and the "correct" choice of C). An equation for this singularity is $z_1^p + z_2^q + z_3^r + z_1 z_2 z_3$; we denote its isomorphism class by $T_{p,q,r}$. Karras (op. cit.) also determines the cases for which we get a complete intersection: this is so if and only if $\sigma^{\pm 1}$ is conjugate to

$$\begin{pmatrix} 0 & 1 \\ -1 & p \end{pmatrix}\begin{pmatrix} 0 & 1 \\ -1 & r \end{pmatrix}\begin{pmatrix} 0 & 1 \\ -1 & q \end{pmatrix}\begin{pmatrix} 0 & 1 \\ -1 & s \end{pmatrix}$$

for certain positive integers p,q,r,s with $(\frac{1}{p}+\frac{1}{q})(\frac{1}{r}+\frac{1}{s}) < 1$. Equations are

$$z_1^p+z_2^q = z_3z_4$$
$$z_3^r+z_4^s = z_1z_2.$$

We denote its isomorphism class by $T_{p,q,r,s}$ (one may check that $T_{p,q,r,1} \cong T_{p,q,r+1}$).

*We close this section by sketching an alternate construction of the triangle singularities based on the realization of $PSL_2(R)$ as the identity component of $SO(2,1)$, which is in some respects a more natural way to introduce them. Actually, the construction yields a one parameter family $\{(X_t,0);t \in C\}$ of singularities: for $t=0$ we get the triangle singularity, but for $t \neq 0$, we find a different isomorphism type. As before, we let $p \leq q \leq r$ be positive integers with $\frac{1}{p}+\frac{1}{q}+\frac{1}{r} < 1$. We endow R^3 with the inner product (.) whose matrix is given by

$$\begin{bmatrix} 1 & -\cos\frac{\pi}{r} & -\cos\frac{\pi}{q} \\ -\cos\frac{\pi}{r} & 1 & -\cos\frac{\pi}{p} \\ -\cos\frac{\pi}{q} & -\cos\frac{\pi}{p} & 1 \end{bmatrix}$$

One easily checks that (.) is nondegenerate and has signature (2,1). So if G denotes the orthogonal group of (.), then $G \cong SO(2,1)$. Consider the subset of C^3 defined by the equality $(\omega.\omega) = 0$ and the inequality $(\omega.\overline{\omega}) > 0$. If we write $\omega = x+iy$, we see that this means that $(x.x) = (y.y) > 0$ and $(x.y) = 0$. From this it is not hard to see that the subset in question has two connected components (interchanged by complex conjugation). Let Ω_0 be one of these components. Then Ω_0 is left invariant by the identity component G_0 of G and is in fact a principal homogeneous space for it. Moreover, the G-equivalent quotient map $\Omega_0 \to \mathbf{P}(\Omega_0)$ can be identified with the $PSL_2(\mathbf{R})$ projection $T^*H-H \to H$. Let $s_i \in G$ denote the reflection in $\mathbf{R}^3$ orthogonal to the basis vector e_i (with respect to (.)): $s_i(x) = x-2(x.e_i)e_i$. One checks that s_2s_3, s_3s_1, s_1s_2 have order p,q,r respectively and are in G_0. The subgroup Γ_+ of G_0 generated by these elements is isomorphic to the triangle group $\Sigma_+ \subset PSL_2(\mathbf{R})$ defined in ex. 6. So by adding a singleton $\{\infty\}$ to $\Gamma_+\backslash\Omega_0$ we obtain an affine surface $\widehat{\Gamma_+\backslash\Omega_0}$ with a triangle singularity at ∞. The coordinate ring of this surface can be obtained as follows. Let A_0^p denote the space of Γ_+-invariant analytic functions on Ω_0 which are homogeneous of degree p (i.e. satisfy $\phi(\lambda\omega) = \lambda^p\phi(\omega)$). Then it turns out that $A_0^p = 0$ for $p > 0$, $A_0^0 = \mathbf{C}$ and the $\mathbf{C}$-algebra $A_0^* := \oplus_{p\in\mathbf{Z}} A_0^p$ is noetherian. This $\mathbf{C}$-algebra describes $\widehat{\Gamma_+\backslash\Omega}$ (with a $\mathbf{C}^*$-action inherited from scalar multiplication in $\mathbf{C}^3$) and the maximal ideal $\oplus_{p<0} A_0^p$ defines ∞.

For $\omega = x+iy \in \mathbf{C}^3$, the inequality $|(\omega.\omega)| < (\omega.\bar{\omega})$ means that x and y span a positive definite plane in $\mathbf{R}^3$. From this it easily follows that the (open) subset of $\mathbf{C}^3$ defined by this inequality has two connected components. Let Ω be the component which contains Ω_0. The group G_0 leaves Ω invariant, G_0 acts properly and freely on Ω and the orbits are just the fibres of the G_0-invariant function $f : \Omega \to \mathbf{C}$, $f(\omega) = (\omega.\omega)$. (Notice that $\Omega_0 = f^{-1}(0)$). Let A^p denote the space of Γ_+-invariant analytic functions on Ω homogeneous of degree p and put $A^* = \oplus_{p\in\mathbf{Z}} A^p$. An essential difference with the previous case is that the A^p may be nonzero for $p > 0$ (in fact $f \in A^2$) and may have infinite $\mathbf{C}$-dimension. Nevertheless, we have $A_0^*/fA_0^* \cong A_0$ and if we put $I := \oplus_{p<0} A^p$ then the natural map $\mathbf{C}[f] \to A^*/I$ is an isomorphism. The $\mathbf{C}$-algebra A^* is the algebra of analytic functions or a (Stein) space $\widehat{\Gamma_+\backslash\Omega}$ obtained by adding to $\Gamma_+\backslash\Omega$ a copy S of $\mathbf{C}$. The ideal defining S in $\widehat{\Gamma_+\backslash\Omega}$ is I so that the coordinate ring of

S may be identified with $\mathbf{C}[f]$. We have a $\mathbf{C}^*$-action on $\Gamma_+\backslash\Omega$ induced by the $\mathbf{C}^*$-action on Ω, which we extend to $\widehat{\Gamma_+\backslash\Omega}$ by letting $\mathbf{C}^*$ act on $S \cong \mathbf{C}$ by $\lambda.s = \lambda^2 s$. The Γ_+-invariant function f factors over a function $\pi : \Gamma_+\backslash\Omega \to \mathbf{C}$ which extends to $\hat{\pi} : \widehat{\Gamma_+\backslash\Omega} \to \mathbf{C}$. This $\hat{\pi}$ will map S isomorphically to $\mathbf{C}$ and if we denote by $s_t \in S$ the unique point over $t \in \mathbf{C}$, then s_t is the unique singular point of the fibre $\hat{\pi}^{-1}(t)$. The pair $(\hat{\pi}^{-1}(0),s_0)$ can be identified with the pair $(\widehat{\Gamma_+\backslash\Omega_0},\infty)$ and hence defines a triangle singularity of type (p,q,r). The germs $\{(\hat{\pi}^{-1}(t),s_t)\}_{t\in\mathbf{C}-\{0\}}$ are mutually isomorphic since $\mathbf{C}^*$ is transitive on $S-\{s_0\}$. Each of them is not analytically isomorphic to $(\hat{\pi}^{-1}(0),s_0)$, although very similar to it (the germs in question are topologically equivalent). It can be shown that for $t \neq 0$, $(\hat{\pi}^{-1}(t),s_t)$ is not analytically equivalent to a cone singularity; for that reason we refer to its isomorphism class as the *non quasi-homogeneous* (companion of the) *triangle singularity*.

For instance, if we take (p,q,r) = (3,3,4), then let $X \subset \mathbf{C}^4$ be the affine variety defined by the equation $x^2+y^3+z^7+syz^5$. The $\mathbf{C}^*$-action on $\mathbf{C}^4$ defined by $\lambda.(x,y,z,s) = (\lambda^{21}x,\lambda^{14}y,\lambda^6 z,\lambda^{-2}s)$ leaves X invariant. Then $\widehat{\Gamma_+\backslash\Omega}$ can be identified with a $\mathbf{C}^*$-invariant open neighbourhood of $0 \in \mathbf{C}^4$ and this identification makes S respectively $\hat{\pi}$ correspond to the s-axis respectively the last coordinate.

2 THE MILNOR FIBRATION

Although the objects studied in this book are mostly germs of spaces and of mappings, many of the results we are going to discuss concern *representatives* of germs. It is then convenient to have a distinguished class of representatives whose members are sufficiently alike (for instance, are mutually topologically equivalent). In this chapter we describe such classes: in §A we do this for an isolated analytic singula

and in §B we generalize this to families of such singularities. This leads to the notion of a 'good representative'. The next section concentrates on the geometric monodromy representation. In §D we discuss an even better (hence smaller) class of representatives: the excellent ones. This section stands somewhat apart, in that we state results whose proofs are merely sketched (and presuppose some knowledge of stratification theory). The reason is that although we don't make use of these facts, it is good to be aware of them.

2.A *The link of an isolated singularity*

In this section X is an analytic set in an open $U \subset \mathbf{C}^N$ and $x \in X$ is such that $X-\{x\}$ is nonsingular of constant dimension n. The main result will be that at x, X is homeomorphic to a cone over a C^∞-manifold and that this manifold is unique up to diffeomorphism.

We begin with citing the

(2.1) *Curve Selection Lemma*. Let V be an open neighbourhood of $p \in \mathbb{R}^m$ and let $f_1,\dots,f_k$, $g_1,\dots,g_\ell$ be real-analytic functions on V such that p is in the closure of $Z := \{x \in V : f_\kappa(y) = 0\ \kappa=1,\dots,k;\ g_\lambda(y) > 0\ \lambda=1,\dots,\ell\}$. Then there exists a real-analytic curve $\gamma : [0,\delta) \to V$ with $\gamma(0) = p$ and $\gamma(t) \in Z$ for $t \in (0,\delta)$.

We shall not prove this here, but refer to Milnor (1968), Lemma (3.1). He assumes $V = \mathbb{R}^m$ and the f_κ, g_λ polynomials, but his proof also works in the present analytic context (with only minor modifications).
As an application we have the following useful result.

(2.2) *Lemma*. Let $r : X \to [0,\infty)$ be the restriction of a real-analytic function $\tilde{r}$ defined on a neighbourhood of X in U such that $r^{-1}(0) = \{x\}$. Then 0 is not an accumulation point of critical values of $r|X-\{x\}$.
Proof. The question being of a local nature, we may assume that (by possibly shrinking U) there exist finitely many holomorphic functions $f_1,\dots,f_k$ on U which generate I_X. So for all $y \in X-\{x\}$, the system $df_1(y),\dots,df_k(y)$ has rank $N-n$. Let Y denote the set of $y \in X$ where $\{d\tilde{r}(y), df_1(y),\dots,df_k(y), d\bar{f}_1(y),\dots,d\bar{f}_k(y)\}$ has rank $\le 2(N-n)$. Clearly, Y is a real-analytic subset of X which contains x and $Y-\{x\}$ is just the critical set of $r|X-\{x\}$. It suffices to show that x is an isolated point of Y. Suppose not: then by the Curve Selection Lemma we can find a real-analytic curve $\gamma : [0,\delta) \to \mathbb{C}^N$ with $\gamma(0) = x$ and $\gamma(t) \in Y-\{x\}$ if $t \neq 0$. Since $(r\circ\gamma)^\cdot(t) = \langle dr(\gamma(t)), \dot{\gamma}(t)\rangle = 0$, it follows that $r\circ\gamma(t) = r\circ\gamma(0) = 0$. Hence $\gamma(t) = x$ for all t, contradicting our assumption that $\gamma(t) \neq x$ for $t \neq 0$.

(2.3) A function r as in the lemma, is said to *define the point* x *in* X. We use the following notation: $X_{r\leq\varepsilon} = \{x \in X : r(x) \leq \varepsilon\}$ and similarly, $X_{r<\varepsilon}$, $X_{r=\varepsilon}$, $X_{0<r<\varepsilon}$, ...

Recall that the *cone* over a space Z, Cone(Z), is obtained from $[0,1] \times Z$ by collapsing $\{0\} \times Z$ to a point.

(2.4) *Proposition.* Let $r : X \to [0,\infty)$ define x in X and let $\varepsilon > 0$ be such that $X_{r\leq\varepsilon}$ is compact and $r|X-\{x\}$ has no critical value in $(0,\varepsilon]$. Then $X_{r=\varepsilon}$ is a compact real-analytic submanifold of X and there exists a homeomorphism H of the cone of $X_{r=\varepsilon}$ onto $X_{r\leq\varepsilon}$ such that $\frac{1}{\varepsilon} r \circ H$ is the projection onto $[0,1]$.

Proof. That $X_{r=\varepsilon}$ is a real-analytic submanifold of X follows from the implicit function theorem. Clearly, $X_{r=\varepsilon}$ is compact. Construct a C^∞-vector field v on $X_{0<r\leq\varepsilon}$ (tangent to X) with $\langle dr,v\rangle = -\varepsilon$. It is not difficult to find such a vector field locally; a global one is obtained from the local solutions by means of a partition of unity. For $y \in X_{0<r\leq\varepsilon}$, let $h_y : (-\delta,\delta) \to X-\{0\}$ denote the integral curve of v with $h_y(0) = y$. Then $(r\circ h_y)^\cdot(t) = \langle dr(h_y(t)),\dot{h}_y(t)\rangle = \langle dr(h_y(t)),v(h_y(t))\rangle = -\varepsilon$, so that $r\circ h_y(t) = r(y)-\varepsilon t$. Since $r : X_{r\leq\varepsilon} \to [0,\varepsilon]$ is proper with $r^{-1}(0) = \{x\}$, it follows that h_y is defined on $[0,\varepsilon^{-1}r(y))$ with $h_y(t) \to x$ as $t \to \varepsilon^{-1}r(y)$. The uniqueness theorem for ODE's implies that the map $H : (0,1] \times X_{r=\varepsilon} \to X_{0<r\leq\varepsilon}$, $H(t,y) = h_y(\varepsilon(1-t))$ is a diffeomorphism with $r\circ H(t,y) = \varepsilon t$. So H extends to a homeomorphism of $\mathrm{Cone}(X_{r=\varepsilon})$ onto $X_{r\leq\varepsilon}$ with the desired property.

We shall call the submanifold $X_{r=\varepsilon}$ of X a *link* of x in X. The following proposition implies that two such are diffeomorphic.

(2.5) *Proposition.* Let $r,r' : X \to [0,\infty)$ define x in X. Then there exists an $\varepsilon > 0$ such that

(i) the hypotheses of (2.4) are satisfied for r and ε.

(ii) if $\varepsilon' > 0$ is such that $X_{r'\leq\varepsilon'} \subset X_{r<\varepsilon}$, then the hypotheses of (2.4) are satisfied for r' and ε' and there exists a diffeomorphism of $X_{r\leq\varepsilon,r'\geq\varepsilon'}$ onto $[0,1] \times X_{r=\varepsilon}$ which maps $X_{r=\varepsilon}$ respectively $X_{r'=\varepsilon'}$ onto $\{0\} \times X_{r=\varepsilon}$ respectively $\{1\} \times X_{r'=\varepsilon'}$.

Proof. This is a modification of the proofs of (2.2) and (2.4). First we claim that there is a neighbourhood V of x in X such that at no $y \in V-\{x\}$, $dr(y)$ and $dr'(y)$ point in opposite directions, that is if $dr(y)$ and $dr'(y)$ are proportional, then the factor of proportionality is never < 0. For let Y be the set of $y \in X$ where $dr(y)$ and $dr'(y)$ are linearly dependent. This is a real-analytic subset of X: if $f_1,\ldots,f_\ell$ are generators of $I_{X,x}$ and r respectively r' are restrictions of real-analytic $\tilde{r}$ respectively $\tilde{r}'$ defined on a neighbourhood of x in X, then near x, Y is the set of $y \in X$ where the system of covectors $\{df_1(y),\ldots,df_\ell(y),d\bar{f}_1(y),\ldots,d\bar{f}_\ell(y), d\tilde{r}(y),d\tilde{r}'(y)\}$ has rank $< 2(N-n+1)$. Let (.) denote the standard euclidean inner product in $\mathbb{C}^N$ and suppose that arbitrarily close to x we can find points of $Y-\{x\}$ where $(dr.dr')$ is < 0. According to the Curve Selection Lemma, there exists then a real-analytic curve $\gamma : [0,\delta) \to Y$ with $\gamma(0) = x$, $(dr.dr')(\gamma(t)) < 0$ for $t > 0$. Hence $dr'(\gamma(t)) = \lambda(t)dr(\gamma(t))$ for some $\lambda(t) < 0$ if $t > 0$. It then follows from $(r'\circ\gamma)^\cdot(t) = \langle dr'(\gamma(t)), \dot{\gamma}(t)\rangle = \lambda(t)\langle dr(\gamma(t)),\dot{\gamma}(t)\rangle = \lambda(t)(r\circ\gamma)^\cdot(t)$, that $r'\circ\gamma$ and $r\circ\gamma$ do not both increase along γ (at x) which is absurd.

Now let $\varepsilon > 0$ be such that $X_{r\leq\varepsilon}$ is compact, contained in V and neither r nor r' has a critical point in $X_{r\leq\varepsilon}$. Using a partition of unity we obtain a C^∞ vector field v on $X_{r\leq\varepsilon}$ such that $\langle dr,v\rangle = -1$, $\langle dr',v\rangle < 0$ and v is tangent to X. If $y \in X_{r=\varepsilon}$, and $h_y : [0,\varepsilon) \to X_{r\leq\varepsilon}$ is the integral curve of v with $h_y(0) = y$, then $r\circ h_y(t) = \varepsilon-t$ and $(r'\circ h_y)^\cdot(t) < 0$. In particular, $r'\circ h_y$ is strictly decreasing and if $\varepsilon' > 0$ is such that $X_{r'\leq\varepsilon'} \subset X_{r\leq\varepsilon}$, then by the implicit function theorem there is a real-analytic $\tau : X_{r=\varepsilon} \to [0,\varepsilon)$ such that $t = \tau(y)$ is the unique solution of $r'\circ h_y(t) = \varepsilon'$. So $G : [0,1]\times X_{r=\varepsilon} \to X_{r\leq\varepsilon, r'\geq\varepsilon'}$, $G(t,y) = h_y(t\tau(y))$ has the required properties.

(2.6) *Corollary*. The diffeomorphism type of a link of x in X only depends on the abstract analytic germ (X,x) (i.e. on the $\mathbb{C}$-algebra $\mathcal{O}_{X,x}$). Indeed, this follows in a straightforward manner from (1.7) and (2.5).

2.B *Good representatives*

(2.7) We now let X be analytic set of pure dim n+k in some open $U \subset \mathbb{C}^N$, $x \in X$ and $F = (F_1,\ldots,F_k) : U \to \mathbb{C}^k$ an analytic map with the property that in each point of $F^{-1}F(x)-\{x\}$, X is nonsingular and $F|X$ is a submersion. As this is clearly a property of the restriction $f := F|X$, we say that then f *defines an isolated singularity at* x (even when f is a submersion in x). For convenience we assume $f(x) = 0$. We also suppose that we are given a real-analytic $r : X \to [0,\infty)$ such that $r|f^{-1}(0)$ defines x in $f^{-1}(0)$. Let $\varepsilon > 0$ be as in (2.4) relative $f^{-1}(0)$. Since $f^{-1}(0)_{r=\varepsilon}$ is compact, this implies that there is a neighbourhood S of 0 in $\mathbb{C}^k$ such that $f|X_{r=\varepsilon}$ is a local submersion along $f^{-1}(S)_{r=\varepsilon}$. We take S contractible. At this point it is convenient to introduce some notation to which we will stick throughout the rest of the book as much as we can. We set

$$X := f^{-1}(S)_{r<\varepsilon} \qquad \bar{X} := f^{-1}(S)_{r\leq\varepsilon} \qquad \partial\bar{X} := f^{-1}(S)_{r=\varepsilon}$$

so that X is open in $f^{-1}(S)$ and $\bar{X}$ resp. $\partial\bar{X}$ is its closure respectively bound-

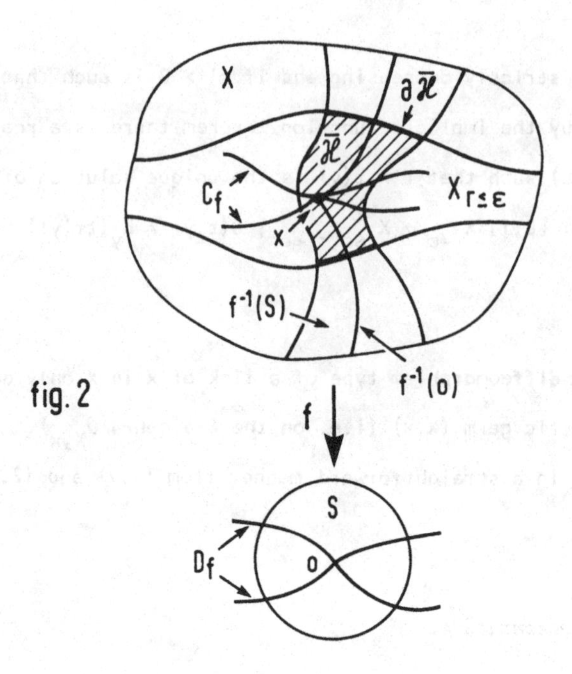

fig. 2

ary in $f^{-1}(S)$. We further let C_f (or C) denote the intersection of X with the *critical locus* of f (i.e. the set of points of X which are singular or where f is not a submersion). If $k = 0$, then C_f is just the singular locus of $\overline{X}$, for which we also use the notation X_{sing}. We write X_{reg} for $X - X_{sing}$. For $s \in S$, let $\overline{X}_s$ respectively X_s denote the intersection of $f^{-1}(S)$ with $\overline{X}$ and X respectively. Similarly, if $A \subset S$, we put $\overline{X}_A := \overline{X} \cap f^{-1}(A)$ and $X_A := X \cap f^{-1}(A)$. The image $f(C_f)$ in S is called the *discriminant (locus)* of f and will be denoted by D_f (or D). We call the restriction $f : X \to S$ respectively $f : \overline{X} \to S$ a *good representative* respectively a *good proper representative* of f.

(2.8) *Theorem.* In this situation we have

(i) $f : \overline{X} \to S$ is proper and $f : \partial\overline{X} \to S$ is a C^∞-trivial fibre bundle

(ii) C_f is analytic in X and closed in $\overline{X}$. Moreover, $f|C_f$ is finite (i.e. proper and with finite fibres).

(iii) The singular locus X_{sing} of X is of dim $\leq k$ and $C_f - X_{sing}$ is of pure dim $k-1$.

(iv) D_f is an analytic subset of S of the same dimension as C_f. It is a hypersurface in S (or void) if $C_f - X_{sing}$ is dense in C_f.

(v) The mapping $f : (\overline{X}_{S-D_f}, \partial\overline{X}_{S-D_f}) \to S-D_f$ is a C^∞-fibre bundle pair, of which each fibre pair $(\overline{X}_s, \partial\overline{X}_s)$ is a complex-analytic n-manifold with boundary.

(vi) f defines an icis at every point of $\overline{X}_{reg}$.

Proof. (i) If $K \subset S$ is compact, then so is $\overline{X} \cap f^{-1}(K)$. Hence $f : \overline{X} \to S$ is proper. For the same reason, $f : \partial\overline{X} \to S$ is proper. As $f|\partial\overline{X}$ is a submersion, the Ehresmann fibration theorem (e.g. Wolf (1964)) tells us that $f|\partial\overline{X}$ is locally trivial. Since S is contractible, $f|\partial\overline{X}$ is in fact globally trivial.

(ii) By construction, f is a submersion at $\partial\overline{X}$, so C is closed in $\overline{X}$. Let $y \in X$. If $g_1, \ldots, g_\ell$ generate $\mathcal{I}_X$ on a neighbourhood V of y in U, then $C \cap V$ is the set of $z \in X \cap V$ where $\{dF_1(z), \ldots, dF_k(z), dg_1(z), \ldots, dg_\ell(z)\}$ has rank $< N-n$. This is clearly an analytic subset of V and so C is analytic in $\overline{X}$. Since C is closed in $\overline{X}$, the restriction $f : C \to S$ is proper. In particular, $f^{-1}(s) \cap C$ is a compact analytic subset of X and hence finite.

(iii), (iv). The fact that $f : C \to S$ is finite implies that $D = f(C)$ is an analytic subset of S of the same dimension as C (by the finite mapping theorem). In particular, $\dim C \leq k$. If y is a regular point of X, then $f|C$ is not open in y (by Sard's theorem) and so $\dim(C,y) \leq k-1$. To see that $\dim(C,y) \geq k-1$ also, we choose local coordinates for X in y, so that the jacobian matrix of f in y is given by a map-germ $df : (X,y) \to \mathrm{Hom}(\mathbb{C}^{n+k}, \mathbb{C}^k)$. Let $\Sigma \subset \mathrm{Hom}(\mathbb{C}^{n+k}, \mathbb{C}^k)$ denote the set of homomorphisms of rank $< k$. This is an affine algebraic variety of pure codimension $n+1$ (this is not hard to check; for a formal proof we refer to Ch. 4, §B). Since the germ of C at y is equal to $(df)^{-1}(\Sigma)$ it follows that C is of codim $\leq n+1$ (and hence of dim $\geq k-1$) in y. It is clear that if $C-X_{sing}$ is dense in C, C and hence D will be of pure dimension k-1.

(v) follows from the Ehresmann fibration theorem.

(vi) is immediate from the definition (1.9).

A fibre X_s (respectively $\overline{X}_s$) with $s \in S-D$ is called a (compact) *Milnor fibre*, and the bundle of (2.8.v) (sometimes also the whole map $f : \overline{X} \to S$) is referred to as the *Milnor fibration*, for Milnor (1968) was the first to study this bundle in a systematic fashion in the case k = 1, X is nonsingular. In a sense, the diffeomorphism type of the Milnor fibration only depends on the germ f_x of f at x:

(2.9) *Proposition*. If $\overline{f} : \overline{X} \to S$ and $\overline{f}' : \overline{X}' \to S'$ are good proper representatives of f_x then there exist a neighbourhood T of 0 in $S \cap S'$ and a C^∞-diffeomorphism $H : (\overline{X}_T, \partial\overline{X}_T) \to (\overline{X}'_T, \partial\overline{X}'_T)$ which is the identity on a neighbourhood of $C_f \cap f^{-1}(T)$ and commutes with the projection onto T. In particular, H induces a diffeomorphism $\overline{X}_s \to \overline{X}'_s$ for all $s \in T$.

Proof. Choose a good proper representative $f : \overline{X}'' \to S''$ with $\overline{X}'' \subset X \cap X'$ (and hence $S'' \subset S \cap S'$). Since f is a submersion on the compact set $\overline{X}_0 - X''_0$, there exists an open contractible neighbourhood T of 0 in $S \cap S'$, such that f is a submersion on $\overline{X}_T - X''_T$. Now $f : \overline{X}_T - X''_T \to T$ is proper and its restriction to the boundary $\partial\overline{X}_T \cup \partial\overline{X}''_T$ is also a submersion. Following Ehresmann, $f : (\overline{X}_T - X''_T; \partial\overline{X}''_T, \partial\overline{X}_T) \to T$ is then a fibre bundle triple. Because T is contractible, this bundle is trivial. According to (2.5.ii) the fibre $(\overline{X}_0 - X''_0; \partial\overline{X}''_0, \partial\overline{X}_0)$ is diffeomorphic to $([0,1];\{0\},\{1\}) \times \partial\overline{X}''_0$. Thus we have a diffeomorphism h of $(\overline{X}_T - X''_T; \partial\overline{X}''_T, \partial\overline{X}_T)$ onto $([0,1];\{0\},\{1\}) \times \partial\overline{X}''_T$ which is the obvious map on $\partial X''_T$ and commutes with the projection onto T. By shrinking T if necessary, we find a similarly defined h' with $\overline{X}_T$ replaced by $\overline{X}'_T$. A standard argument in differential topology (e.g. Munkres (1966), lemma 6.1) makes it possible to choose h' in such a way that $h' \circ h^{-1} : [0,1] \times \partial \overline{X}''_T \circlearrowleft$ is the identity near (instead of on) $\{0\} \times \partial \overline{X}''_T$. If we let H be the identity on $\overline{X}''_T$ and $h'^{-1} \circ h$ on $\overline{X}_T - \overline{X}''_T$, H is as desired.

At this point it is not hard to prove the following lemma, although we won't need it until Ch. 5.

(2.10) *Lemma*. Suppose that in the situation of (2.8), X-{x} is nonsingular. Then for any good proper representative $f : \overline{X} \to S$ of f, there exist an $\eta_0 > 0$ such that for any $\eta \in (0,\eta_0]$, $\overline{X}_{|f| \leq \eta}$ is homeomorphic to the cone on its boundary $\partial \overline{X}_{|f| \leq \eta} \cup \overline{X}_{|f|=\eta}$.

Proof. In view of (2.9) we only need to prove this for one good representative. In particular, we may assume that r defines x in X. We claim that there exist $\varepsilon > 0$, $\eta_0 > 0$ such that the differentials of r and $|f|^2 = |f_1|^2 + \ldots + |f_k|^2$ do not point in opposite directions in y if $r(y) \in (0,\varepsilon]$, $|f(y)| \in (0,\eta]$. This easily follows from the Curve Selection Lemma. By means of a partition of unity we are then able to find a vector field v on $X_{r\leq\varepsilon, |f| \leq \eta_0} - \{x\}$ with $\langle dr, v \rangle = -1$ and $\langle d|f|^2, v \rangle \leq 0$. This vector field has the property that its integral curves can be extended as to end in x and that $|f|^2$ nowhere increases along such a curve. It follows that this describes a contraction of $\overline{X}_{|f|\leq\eta} = X_{r\leq\varepsilon, |f|\leq\eta}$ onto {x}.

We mention a particular property the Milnor fibres have when X is nonsingular.

(2.11) *Proposition*. If $f : X \to S$ is a good representative with X nonsingular then the tangent bundle of any Milnor fibre X_s is trivial as a complex vector bundle.

Proof. We may assume that X is small enough to be contractible. Then its complex tangent bundle will be trivial. The differentials $df_1,\ldots,df_k$ trivialize the normal bundle of X_s in X and so the complex tangent bundle of X_s is stably trivial. But this is equivalent to the bundle being actually trivial, see Husemoller (1966), Ch. 8, Thm (1.5).

The case originally considered by Milnor (1968) is k = 1, X = U open in $\mathbf{C}^{n+1}$, for which he used a somewhat different approach. Instead of taking a good representative, Milnor considers the map

$$\arg(f) = f/|f| : S_\varepsilon-(X_0 \cap S_\varepsilon) \to S^1,$$

where $S_\varepsilon = \{y \in \mathbf{C}^{n+1} : |y-x| = \varepsilon\}$, and proves that for sufficiently small $\varepsilon > 0$, arg(f) is a C^∞-fibre bundle. Each fibre F_θ is an open 2n-submanifold of S_ε which admits $X_0 \cap S_\varepsilon$ as its boundary and $\overline{F}_\theta$ is diffeomorphic to a compact Milnor fibre of f. An advantage of this approach is that it leads to valuable information about the way $X_0 \cap S_\varepsilon$ is embedded in S_ε. It also enables Milnor to give a quick proof that the Milnor fibre has the homotopy type of a bouquet of n-spheres (a result which also holds in the complete intersection case and which we will prove in Ch. 5).

2.C *Geometric monodromy*

Let $f : \overline{X} \to S$ be a good proper representative as in (2.7). In order to investigate the Milnor fibration $f : \overline{X}_{S-D} \to S-D$ we fix a trivialization $\sigma : \partial\overline{X} \to \partial\overline{X}_0$ of $f|\partial\overline{X}$ and choose a base point $s_0 \in S-D$. Since f is locally trivial over S-D there exists for each $s \in S-D$ an open neighbourhood V_s of s in S-D and a trivialization $u_s : \overline{X}_{V_s} \to \overline{X}_s$ compatible with $\sigma : \sigma|\partial\overline{X}_{V_s} = \sigma \circ u_s|\partial\overline{X}_{V_s}$. Let $\gamma : [0,1] \to$ S-D be a (continuous) loop emanating from s_0. As [0,1] is compact, there exists a partition $0 = t_0 < t_1 < \ldots < t_M = 1$ of [0,1] such that γ maps each segment $[t_{\mu-1}, t_\mu]$ to some V_s. By using our trivializations, we find a family of diffeomorphisms of pairs:

$$h_t : (\overline{X}_{s_0}, \partial\overline{X}_{s_0}) \to (\overline{X}_{\gamma(t)}, \partial\overline{X}_{\gamma(t)}),\ t \in [0,1]$$

with $\sigma \circ h_t|\partial\overline{X}_{s_0} = \sigma|\partial\overline{X}_{\gamma(t)}, h_0 = 1$ which is continuous in the sense that $[0,1]\times\overline{X}_{s_0} \to \overline{X}$, $(t,y) \to h_t(y)$ is so. In particular, the self-diffeomorphism h_1 of $\overline{X}_{s_0}$ satisfies $h_1|\partial\overline{X}_{s_0} = 1$. It may depend on choices made, but the relative isotopy class of h_1 does not. In fact, we have the stronger assertion that the relative isotopy class of h_1 only depends on the homotopy class $[\gamma]$ of γ in $\pi(\text{S-D}, s_0)$. (The argument is standard and therefore only outlined: if $\Gamma : [0,1]\times[0,1] \to$ S-D is a homotopy from γ to some γ', then partition $[0,1]\times[0,1]$ into squares, each of which is mapped by Γ to some V_s. Find inductively (using the trivializations u_s) a continuous family of diffeo's $h_{t',t} : \overline{X}_{s_0} \to \overline{X}_{\Gamma(t',t)}$ with $h_{0,t}$ and $h_{1,t}$ as given, $h_{t',0} = 1$. Then $h_{t',1}$ is an isotopy between $h_{0,1}$ and $h_{1,1}$.) We then say that h_1 (or rather its relative isotopy class) is the C^∞-*monodromy* of f along $[\gamma]$. Denoting the group of relative isotopy classes of C^∞-diffeomorphisms of $(\overline{X}_{s_0}, \partial\overline{X}_{s_0})$ by $\mathrm{Iso}^\infty(\overline{X}_{s_0}, \partial\overline{X}_{s_0})$, the resulting map

$$\rho_{diff} : \pi(\text{S-D}, s_0) \to \mathrm{Iso}^\infty(\overline{X}_{s_0}, \partial\overline{X}_{s_0})$$

is clearly a homomorphism of groups. We call ρ_{diff} the C^∞-*monodromy representation* of $\pi(\text{S-D}, s_0)$. Likewise, we have the C^0-(or *geometric*) *monodromy representation*

$$\rho_{geom} : \pi(\text{S-D}, s_0) \to \mathrm{Iso}^0(\overline{X}_{s_0}, \partial\overline{X}_{s_0}).$$

It would be interesting to know to what extent ρ_{geom} determines the Milnor fibration. It presumably does so completely up to topological equivalence if the given trivialization of $\partial\overline{X}_{S-D} \to S\text{-}D$ extends to a trivialization of the pull-back of $\overline{X}_{S-D} \to S\text{-}D$ over a universal covering of S-D. This last condition is fulfilled if there exists a real codim 2n foliation of $\overline{X}$-C which is transversal to the fibres of f and is tangent to $\partial\overline{X}$ (so that $\partial\overline{X}$ is a union of leaves). For k = 1 this has been shown by Deligne in (SGA 7.II), exp. XIV Prop. (3.1.5).

*2.D *Excellent representatives*

In section B we found among other things that two good representatives of the germ f_x have diffeomorphic Milnor fibres. Reassuring as this may be, many properties of the topology of f near x relate to all fibres of f and not just to one. For instance, it is not yet clear that the geometric monodromy group as defined in the previous section has some invariant meaning. We therefore want to single out a class of representatives of f_x, not depending on any coordinate system, such that any two of its members are topologically equivalent (we say that $f : X \to S$ and $f' : X' \to S'$ are topologically equivalent if there exist homeomorphisms $H : X \to X'$ with $H(x) = x$ and $H' : S \to S'$, $H'(0) = 0$ such that $f' = H' \circ f \circ H^{-1}$). Unfortunately, the class of good representatives doesn't possess this property. So, good representatives being not good enough, we need excellent ones. The purpose of this section is to define them and to describe some of their properties, the main one being that these representatives have a conical structure. Our principal tool will be the theory of stratifications, initiated by Thom (1964, 1969) and further developed by Whitney (1965) and Mather (1973). A suitable text book reference is Gibson et al. (1976) Ch. I, II.

Recall that an *analytic stratification* of an analytic set Y is a partition S of Y into connected constructible analytic submanifolds, called *strata*, which is locally finite. It is said to have the *Whitney property* if for each pair of strata $Z_1, Z_2 \in S$, Z_1 is Whitney regular over Z_2. (We refrain from recalling the definition of Whitney regularity here. Simple as that definition may be, for the reader not familiar with stratification theory, it is not very illuminating if one doesn't derive some of its properties.) Now let $f : \overline{X} \to S$ be a good proper representative with critical locus $C \subset X$ and discriminant locus D. Then along the lines of Gibson et al. (1976), (I.3.5), one proves that there exists an analytic Whitney stratification $\mathcal{D}$ of D such that the connected components of C_Z, $X_Z - C_Z$ $(Z \in \mathcal{D})$ define a Whitney stratification E of X_D. We may extend $\mathcal{D}$ and E to (real-analytic) Whitney stratifications of S and $\overline{X}$ by adding to them the strata S-D, and the connected components of X_{S-D}, ∂X_{S-D} and ∂X_Z, $Z \in \mathcal{D}$.

Now let $u : S \to [0,\infty)$ be a real-analytic function which defines $\{0\}$. Precisely as in (2.2) we find an $\eta > 0$ such that $S_{u\leq\eta}$ is a compact subset of S and the restriction of u to any stratum has no critical value in $(0,\eta]$. Then over $S_{u\leq\eta}$, f has a conical structure, by which we mean: $\partial\overline{X}_{u\circ f=\eta}$ admits an open neighbourhood N in $\overline{X}_{u\circ f=\eta}$ and a diffeomorphism of N onto $S_{u=\eta}\times(0,1]\times\partial\overline{X}_0$, such that the map $\rho : \overline{X}_{u\circ f=\eta} \to \mathrm{Cone}(\partial\overline{X}_0)$ which extends the projection of N onto $(0,1]\times\overline{X}_0$ and sends $\overline{X}_{u\circ f=\eta}-N$ onto the vertex has the following property: There exist a homeomorphism H of the mapping cone $\mathrm{Cone}(\rho)$ of ρ onto $\overline{X}_{u\circ f\leq\eta}$ and a homeomorphism H' of $\mathrm{Cone}(S_{u=\eta})$ onto $S_{u\leq\eta}$ which are the obvious maps when restricted to $\overline{X}_{u\circ f=\eta}\times\{1\}$ and $S_{u=\eta}\times\{1\}$ respectively such that the diagram below commutes:

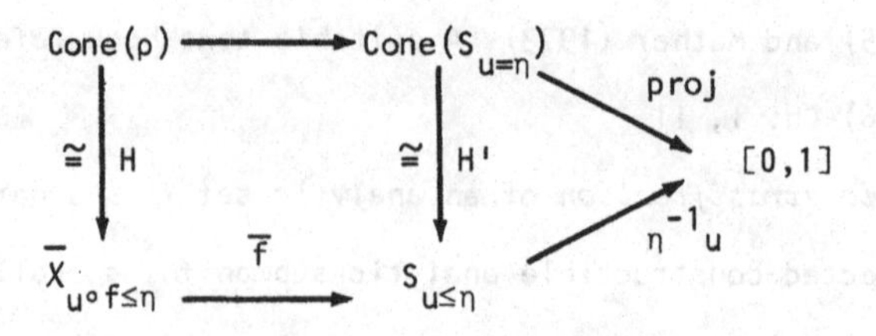

Here the upper horizontal map is induced by $f : \overline{X}_{u\circ f=\eta} \to S_{u=\eta}$. The idea of the proof is to construct first an integrable vector field v' on $S_{u\leq\eta}$ tangent to the strata such that $\langle du,v'\rangle = \eta$ (remember that integrability doesn't imply - and is not implied by - continuity). Then we choose an open neighbourhood W of $\overline{X}_0-\{x\}$ in $\overline{X}-C$ and a C^∞-retraction $\rho : W \to \overline{X}_0-\{x\}$ and we lift v' over f to a vector field v on $\overline{X}$ (this means that $df(v) = v'\circ f$) such that v preserves strata, is integrable and is killed by $d\rho$ on a neighbourhood of $\overline{X}_0-\{x\}$ in W. The homeomorphisms H' and H are obtained by integration of v' and v respectively and N is the set of $y \in \overline{X}_{u=\eta}$ whose integral curves meet $\overline{X}_0-\{x\}$.
If any representative deserves the designation excellent, then so do $f : \overline{X}_{u<\eta} \to S_{u<\eta}$ and its restriction to $X_{u<\eta}$. So we shall call them *excellent proper* (respectively *excellent*) representatives. It can be shown along the lines of (2.5) that any two excellent (proper) representatives are topologically equivalent.

3 PICARD-LEFSCHETZ FORMULAS

(3.0) In 2.C we were dealing with a topological pair (X,A) (namely a compact Milnor fibre and its boundary) and a mapping $h : (X,A) \to (X,A)$ which is the identity on A (a geometric monodromy). The map which assigns to a relative cycle on (X,A) the absolute cycle $h(c)-c$ on X defines the *variation homomorphism*

$$\mathrm{var}_*(h) : H_*(X,A) \to H_*(X),$$

which only depends on the relative homotopy class of h. (If no coefficient group is specified we take integral coefficients.) If $j : (X,\emptyset) \subset (X,A)$ denotes the inclusion, then it is clear that we have a commutative diagram

$$\begin{array}{ccc} H_*(X) & \xrightarrow{h_*-1} & H_*(X) \\ j_* \downarrow & \nearrow^{\mathrm{var}_*(h)} & \downarrow j_* \\ H_*(X,A) & \xrightarrow{h_*-1} & H_*(X,A) \end{array}$$

which shows that $\mathrm{var}_*(h)$ and j_* determine both relative and absolute h_*. This is not the only pleasant property of $\mathrm{var}_*(h)$. Suppose for instance that X is contained in a space Y and that h extends to a mapping $H : Y \to Y$ which is the identity on $B := A \cup (Y-X)$. If $\{X,A\}$ is an excisive couple in Y, then $\mathrm{var}_*(H)$ is simply given by

$$H_*(Y,B) \underset{\cong}{\leftarrow} H_*(X,A) \xrightarrow{\mathrm{var}_*(h)} H_*(X) \to H_*(Y)$$

where the outer maps are the obvious ones. There is no such simple relationship between h_* and H_*.

We want to determine the variation in the simplest non-trivial case, namely of the function $f : \mathbf{C}^{n+1} \to \mathbf{C}$, $f(z) = z_0^2 + \ldots + z_n^2$ at $0 \in \mathbf{C}^{n+1}$. (The hypersurface singularity defined by f is called a *quadratic* singularity.) These are expressed by the Picard-Lefschetz formulas, after Picard who treated in Picard & Simart (1897) the case n = 1 and Lefschetz (1924) who did the general case. The reader will notice that even in this simple case, the explicit description of a geometric monodromy is not so easy. We follow Lamotke's (1975) exposition rather closely.

3.A *Monodromy of a quadratic singularity (local case)*

(3.1) The function $z \in f^{-1}(0)-\{0\} \to |z|$ has no critical points and so a good proper representative $f : \overline{X} \to \overline{\Delta}$ of f is gotten by picking some $\varepsilon > 0$ and putting $\overline{X} = \{z \in \mathbf{C}^{n+1} : |z| \le \varepsilon \text{ and } |f(z)| \le \eta\}$ for some sufficiently small η ($\eta < \frac{1}{2}\varepsilon^2$ is enough) and $\overline{\Delta} = \{t \in \mathbf{C} : |t| \le \eta\}$. Scalar multiplication with $e^{\pi i\theta}$ ($\theta \in \mathbf{R}$) maps $\overline{X}_\eta$ diffeomorphically onto $\overline{X}_{\eta e^{2\pi i\theta}}$. So if we didn't care about the boundary, the involution of $\overline{X}_\eta$ given by $z \to -z$ would represent the monodromy along $\partial\overline{\Delta}$. But we do care and so we choose for a more careful approach.

The fibre $\overline{X}_\eta$ is given by

$$\{x+iy \in \mathbf{C}^{n+1} : |x|^2+|y|^2 \le \varepsilon^2,\ |x|^2-|y|^2 = \eta,\ (x.y) = 0\}.$$

It is convenient to parametrize it with the coordinates

$$u = \frac{x}{|x|},\quad v = \sqrt{\left(\frac{2}{\varepsilon^2-\eta}\right)}y$$

as this identifies $\overline{X}_\eta$ with the unit disc bundle E of the tangent bundle of S^n:

$$E = \{u+iv \in \mathbf{C}^{n+1} : |u|=1,\ |v|\leq 1,\ (u.v)=0\}.$$

The corresponding diffeomorphism $h_0 : E \to \overline{X}_\eta$ is then given by

$$h_0(u+iv) = (\sqrt{\tfrac{1}{2}(\varepsilon^2-\eta)}|v|+\eta)u + i\sqrt{\tfrac{1}{2}(\varepsilon^2-\eta)}v.$$

Now define a one-parameter family of homeomorphisms g_θ, $\theta \in \mathbf{R}$ of E onto itself by letting $g_\theta(u+iv)$ be $e^{-\pi i|v|\theta}(u+i\frac{v}{|v|})$ followed by scalar multiplication of the imaginary part with $|v|$. In coordinates:

$$g_\theta(u+iv) = [u\cos(\pi|v|\theta) + \frac{v}{|v|}\sin(\pi|v|\theta)] +$$

$$+ i[-u|v|\sin(\pi|v|\theta) + v\cos(\pi|v|\theta)].$$

Since sin is an odd function, g_θ is well-defined for $v = 0$. We put

$$h_\theta := e^{\pi i\theta}h_0\circ g_\theta : E \to \overline{X}.$$

Clearly, h_θ is a homeomorphism onto $\overline{X}_{\eta e^{2\pi i\theta}}$. But note that now h_0 and h_1 coincide on ∂E, thus defining a trivialization of $\partial\overline{X}$ over $\partial\overline{\Delta} : \partial\overline{\Delta}\times\partial E \to \partial\overline{X}_{\partial\overline{\Delta}}$, $(\eta e^{2\pi i\theta}, u+iv) \to h_\theta(u+iv)$. This trivialization extends over $\overline{\Delta}$: define $\overline{\Delta}\times\partial E \to \partial\overline{X}$ by

$$(\rho e^{2\pi i\theta}, u+iv) \to e^{\pi i\theta}H_\rho\circ g_\theta,$$

where H_ρ is defined as h_0, but with η replaced by ρ. Notice that this map is also well-defined for $\rho = 0$ (although θ isn't). So the automorphism $h_1\circ h_0^{-1}$ of $\overline{X}_\eta$ represents the geometric monodromy of f along the positively oriented $\partial\overline{\Delta}$. Under the identification of $\overline{X}_\eta$ with E by means of h_0, this corresponds to the automorphism $h_0^{-1}\circ h_1 = h_0^{-1}\circ(-h_0)\circ g_1 = h_0^{-1}\circ h_0\circ -g_1 = -g_1$ of E.

In order to compute the variation of $-g_1$, we first make some observations on the homology of the pair $(E,\partial E)$. The disc bundle $E \to S^n$, $u+iv \to u$, has, of course, its zero section $S^n = \{u+i.0 : |u|=1\}$ as a deformation retract and hence E has the homology of S^n. We orient the unit sphere $S^n \subset \mathbf{R}^{n+1}$ by letting $(e_1,\ldots,e_n)$ be an orientation basis of $T_{e_0}S^n$ (where $(e_0,\ldots,e_n)$ is the standard basis of $\mathbf{R}^{n+1}$). If we regard S^n as the zero section of E, then the fundamental class $[S^n]$ is a generator of $H_n(E)$. We orient E via h_0 by the orientation on $\overline{X}_n$ coming from the complex structure. So an oriented coordinate basis of $T_{e_0}(E)$ is given by $(e_1,ie_1,e_2,ie_2,\ldots,e_n,ie_n)$. Given this orientation, Lefschetz duality identifies $H_n(E,\partial E)$ with $H^n(E)$ and by the universal coefficient formula, the latter is just the $\mathbf{Z}$-dual of $H_n(E)$. The corresponding perfect pairing

$$<\ ,\ > : H_n(E,\partial E) \times H_n(E) \to \mathbf{Z}$$

is the intersection pairing. We let $D^n \subset E$ denote the fibre over u^0: $D^n = \{u^0+iv : v_0=0,\ |v|\le 1\}$, which we orient by $(ie_1,\ldots,ie_n)$. Then its fundamental class $[D^n,\partial D^n]$ represents a generator of $H_n(E,\partial E)$ and $<[D^n,\partial D^n],[S^n]>$ is the sign of the permutation $(ie_1,\ldots,ie_n,e_1,\ldots,e_n) \to (e_1,ie_1,\ldots,e_n,ie_n)$, which is $(-1)^{\frac{1}{2}n(n+1)}$. Together with the natural map $j_* : H_n(E) \to H_n(E,\partial E)$, the intersection pairing gives the intersection product

$$(\ .\ ) : H_n(E)\times H_n(E) \to \mathbf{Z},\ (\alpha.\beta) = <j_*(\alpha),\beta>.$$

As the unit disc bundle of the tangent bundle of S^n, E acquires an orientation, which in u^0 is given by $(e_1,\ldots,e_n,ie_1,\ldots,ie_n)$. With respect to this orientation, $([S^n].[S^n])$ is the euler number $1+(-1)^n$ of S^n. So if we use the complex orientation, then we must multiply this with the signature of the permutation $(e_1,\ldots,e_n,ie_1,\ldots,ie_n) \to (e_1,ie_1,\ldots,e_n,ie_n)$, which is $(-1)^{\frac{1}{2}n(n-1)}$.

Now let us see what $var_*(-g_1)$ does to $[D^n, \partial D^n]$. Since the projection of E onto S^n, $u+iv \to u$ is a homotopy inverse to the zero section, we only need to consider the first coordinate of $-g_1$. This is the map $\psi : (D^n, \partial D^n) \to (S^n, \{u^0\})$,

$$\psi(u^0+iv) = -u^0 \cos(\pi|v|) - \frac{v}{|v|}\sin(\pi|v|).$$

This is an identification mapping: it maps $\mathring{D}^n$ homeomorphically onto $S^n - \{u^0\}$ and ∂D^n onto $\{u^0\}$. By construction, $var_*(-g_1)$ sends $[D^n, \partial D^n]$ to $\deg(\psi).[S^n]$. Now it is easily seen that $-\psi$ leaves u^0 fixed and is differentiable at u^0 with derivative the identity. So $\deg(-\psi) = 1$ and hence $\deg(\psi) = (-1)^{n+1}$. So if we let $\delta \in H_n(\overline{X}_\eta)$ be any generator (say the image of $[S^n]$ under h_{0*}) and if $\delta^* \in H_n(\overline{X}_\eta, \partial\overline{X}_\eta)$ is the unique element with $\langle\delta^*, \delta\rangle = 1$, then it follows from the preceding that

(3.1.a) $\quad var_*(h)(\delta^*) = (-1)^{n+1}.(-1)^{\frac{1}{2}n(n+1)}\delta = -(-1)^{\frac{1}{2}n(n-1)}\delta$

(3.1.b) $\quad (\delta.\delta) = (-1)^{\frac{1}{2}n(n-1)} + (-1)^{\frac{1}{2}n(n+1)}$

and so

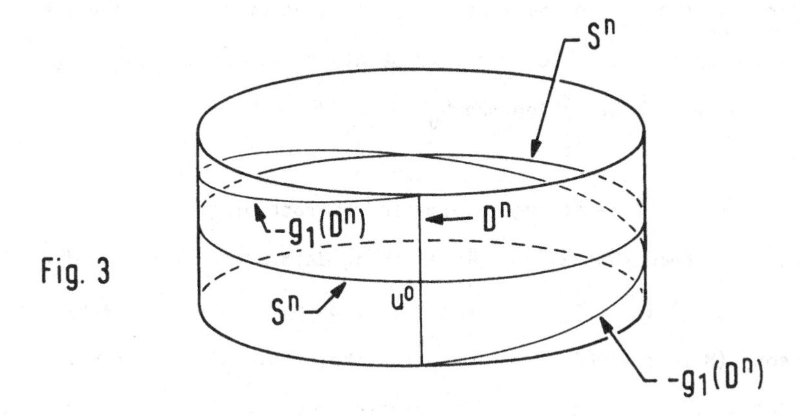

Fig. 3

(3.1.c) $\quad j_*(\delta) = [(-1)^{\frac{1}{2}n(n-1)}+(-1)^{\frac{1}{2}n(n+1)}]\delta^*.$

Since δ^* generates $H_n(\overline{X}_\eta,\partial\overline{X}_\eta)$, the first equality can also be written

(3.1.d) $\quad var_*(h)(c) = -(-1)^{\frac{1}{2}n(n-1)}\langle c,\delta\rangle\delta,\ c \in H_*(\overline{X}_\eta,\partial\overline{X}_\eta).$

The formulas (a), (b) and their equivalent forms (c), (d) are known as the *Picard-Lefschetz formulas*.

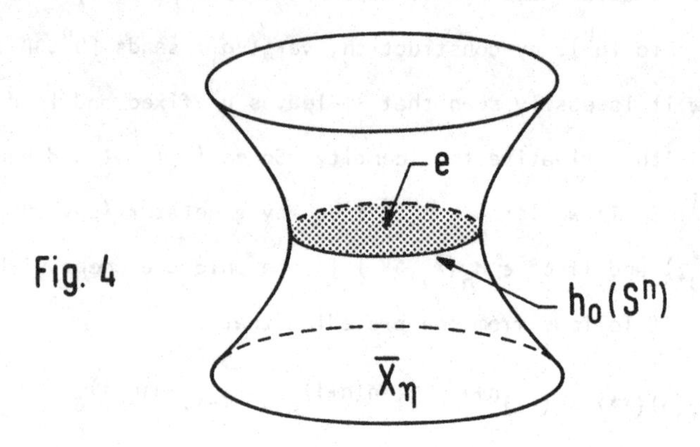

Before passing to a global context, we show that from the homotopy point of view $\overline{X}_0$ is obtained from $\overline{X}_\eta$ by attaching an (n+1)-cell to it. Let e be the closed (n+1)-cell in $\overline{X}$ defined by $|x|^2 \le \eta$ and $y = 0$. Then $\overline{X}_\eta \cap e = \partial e = h_0(S^n)$. So $\overline{X}_\eta \cup e$ is obtained from $\overline{X}_\eta$ by attaching an (n+1)-cell to it with attaching map $h_0|S^n$.

(3.2) *Lemma*. $\overline{X}_\eta \cup e$ is a strong deformation retract of $\overline{X}$.

Proof. We first show that $\overline{X}_{[-\eta,\eta]}$ is a strong deformation retract of $\overline{X}$. The function $g = \mathrm{Im}(f) : \overline{X} \to \mathbb{R}$ is a Morse function. A standard lemma from Morse theory (Milnor (1963) I.3.4) implies that there exists a $\delta > 0$ such that $g^{-1}(0) = \overline{X}_{[-\eta,\eta]}$ is a strong deformation retract of $g^{-1}([-\delta,\delta])$. Now $\overline{f} : \overline{X} \to S$ is a trivial bundle over $\overline{\Delta}-[-\eta,\eta]$ and so the assertion follows.

It therefore remains to prove that $\overline{X}_\eta \cup e$ is a strong deformation retract of $\overline{X}_{[-\eta,\eta]}$.

Consider the map $[0,\eta]\times E \to \overline{X}_{[0,\eta]}$ defined by

$$(\rho, u+iv) \to H_\rho(u+iv) = (\sqrt{\tfrac{1}{2}(\varepsilon^2-\rho)}\,|v|+\rho)u + i\sqrt{\tfrac{1}{2}(\varepsilon^2-\rho)}\,v.$$

It sends $\{0\}\times S^n$ to $\{0\}$ and maps the complement bijectively onto $\overline{X}_{[0,\eta]}-\{0\}$. So this induces a homeomorphism of the quotient space $([0,\eta]\times E)/\{0\}\times E$ onto $\overline{X}_{[0,\eta]}$. In a similar way we find such a homeomorphism over $[-\eta,0]$ and together they make up a homeomorphism

$$([-\eta,\eta]\times E)/\{0\}\times S^n \to \overline{X}_{[-\eta,\eta]}.$$

Since $\{\eta\}\times E \cup ([0,\eta]\times S^n)/\{0\}\times S^n$ is a strong deformation retract of the left hand side, its image, $\overline{X}_\eta \cup e$ is a strong deformation retract of the right hand side.

3.B *Monodromy of a quadratic singularity (global case)*

(3.3) Suppose that the situation investigated in A is part of a global one: we are given a complex $(n+1)$-manifold $\overline{\mathcal{Y}}$ with boundary $\partial\mathcal{Y}$, a point x in its interior $\mathcal{Y}$ and a proper holomorphic map F from $\mathcal{Y}$ to $\Delta = \{t \in \mathbf{C} : |t|<1\}$ with $F(x) = 0$. We assume that $F|\partial\mathcal{Y}$ and $F|\mathcal{Y}-\{x\}$ are submersions and that at x there exist complex coordinates $z_0,\dots,z_n$ such that $F(z) = z_0^2+\dots+z_n^2$. Then $F : (\overline{\mathcal{Y}}-\overline{Y}_0, \partial\overline{\mathcal{Y}}-\partial\overline{Y}_0)) \to \Delta-\{0\}$ is a fibre bundle pair. If we take $\eta \in (0,1]$ sufficiently small, then we can assume that Y_η contains a compact Milnor fibre $\overline{X}_\eta$ of the germ of F at x. The Ehresmann fibration theorem implies that a geometric monodromy $\tilde{h}$ of the loop $\theta \in [0,1] \to e^{2\pi i\theta}\eta$ can be taken to be h (as found in A) on $\overline{X}_\eta$ and the identity on $\overline{Y}_\eta - \overline{X}_\eta$. Then the variation of $\widetilde{h} : (\overline{Y}_\eta, \partial\overline{Y}_\eta)\circlearrowleft$ is the composition

$$\begin{array}{ccc} H_*(\overline{Y}_\eta, \partial\overline{Y}_\eta) \to H_*(\overline{Y}_\eta, \overline{Y}_\eta - X_\eta) & & \\ \uparrow\cong & & \\ H_*(\overline{X}_\eta, \partial\overline{X}_\eta) & \xrightarrow{var_*(h)} & H_*(\overline{X}_\eta) \xrightarrow{i_*} H_*(\overline{Y}_\eta), \end{array}$$

where all maps except the third are induced by inclusions. So if we put $\widetilde{\delta} := i_*(\delta) \in H_n(\overline{Y}_\eta)$, the Picard-Lefschetz formulas give

$$(3.3.a) \qquad (\widetilde{\delta}.\widetilde{\delta}) = (-1)^{\frac{1}{2}n(n-1)} + (-1)^{\frac{1}{2}n(n+1)}$$

$$(3.3.b) \qquad var_*(\widetilde{h})(c) = -(-1)^{\frac{1}{2}n(n-1)} \langle c, \widetilde{\delta}\rangle \widetilde{\delta}, \qquad c \in H_*(\overline{Y}_\eta, \partial\overline{Y}_\eta).$$

In particular, we find that for $\alpha \in H_*(\overline{Y}_\eta)$

$$(3.3.c) \qquad \widetilde{h}_*(\alpha) = \alpha - (-1)^{\frac{1}{2}n(n-1)} (\alpha.\widetilde{\delta})\widetilde{\delta}.$$

Writing this out according to the residue class of n mod 4 gives

$$\begin{array}{llll} n \equiv 0(4) : & \widetilde{h}_*(v) = v - (v.\widetilde{\delta})\widetilde{\delta}, & (\widetilde{\delta}.\widetilde{\delta}) = 2, & \widetilde{h}_*(\widetilde{\delta}) = -\widetilde{\delta} \\ 1(4) : & \phantom{\widetilde{h}_*(v) = }v - (v.\widetilde{\delta})\widetilde{\delta}, & 0, & \widetilde{\delta} \\ 2(4) : & \phantom{\widetilde{h}_*(v) = }v + (v.\widetilde{\delta})\widetilde{\delta}, & -2, & -\widetilde{\delta} \\ 3(4) : & \phantom{\widetilde{h}_*(v) = }v + (v.\widetilde{\delta})\widetilde{\delta}, & 0, & \widetilde{\delta}. \end{array}$$

The (homological) monodromy $\widetilde{h}_*$ is sometimes called the *Picard-Lefschetz transformation*. Notice, that if n is even, (.) is symmetric and that $\widetilde{h}_*$ is the reflection orthogonal to $\widetilde{\delta}$ (relative (.)). If n is odd, (.) is anti-symmetric and $\widetilde{h}_*$ is then a symplectic transvection. In this case it may happen that $\widetilde{\delta} = 0$, so that $\widetilde{h}$ is the identity. This is for instance the case for the family of curves in $\mathbf{P}^2$ defined by $X_1X_2 = tX_0$: for $t \neq 0$, the corresponding curve is a quadric and hence has trivial first homology group.

Lemma (3.2) and the fact that F is locally trivial over $\Delta-\{0\}$ give:

(3.4) Let $e \subset \mathcal{Y}$ be the (n+1)-cell defined in (3.2). Then $\overline{Y}_\eta \cup e$ is a strong deformation retract of $\overline{\mathcal{Y}}_{|f|\leq\eta}$.

Informally speaking, (3.4) says that from the homotopy point of view the inclusion $\overline{Y}_\eta \subset \mathcal{Y}$ amounts to 'killing' the n-sphere $h_0(S^n)$. This is why $h_0(S^n)$ (endowed with an orientation - there is no preferred one -) is called a *vanishing cycle*. It is a traditional abuse of language to refer to its class $\pm\tilde{\delta}$ in $H_n(\overline{Y}_\eta)$ by the same term.

4 CRITICAL SPACE AND DISCRIMINANT SPACE

The principal aim of this chapter is to supply the critical locus and discriminant locus of a good representative with a structure sheaf giving both the structure of an analytic space. Not surprisingly, these spaces are called critical space and discriminant space respectively. The most important feature of these structure sheaves is that they *'commute with base change'* (a property which sometimes forces these sheaves to contain nilpotent elements). 'Commute with base change' means that if $f : X \to S$ and $f' : X' \to S'$ are good representatives (of possibly different germs) fitting in a cartesian diagram

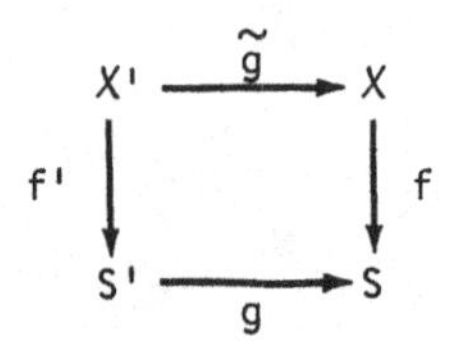

then the ideal sheaf on X' defining the critical space of f' is generated by the image under $\tilde{g}^*$ of the ideal sheaf defining the critical space of f, and similarly for the ideals defining the discriminant spaces. This is a very useful property to have: both for making calculations and for proving things. Following Rim (1972) and Teissier (1972), the structure sheaf for the discriminant is defined with the help of Fitting ideals. For that reason we discuss the notion of a Fitting ideal briefly in section D. Sections B and C are devoted to the Thom stratification of

$\mathrm{Hom}(\mathbb{C}^{n+k},\mathbb{C}^k)$ and its applications. Among other things, it is shown that for most map-germs (namely those which are transversal with respect to the Thom stratification), the map from critical locus to discriminant locus is of degree one and that the so-called Nash-modification of the discriminant is nonsingular.

4.A *The critical space*

Let X be an analytic set in an open $U \subset \mathbb{C}^N$ and let $F : U \to \mathbb{C}^k$ be an analytic map. We put $f : F|X$ and suppose that all fibres of f have the same pure dim n. We then define an ideal sheaf $C_f \subset \mathcal{O}_X$ as follows: if $\phi_1,\ldots,\phi_\ell$ generate $\mathcal{I}_X$ over an open $V \subset U$, then let provisionally $C(\phi_1,\ldots,\phi_\ell)$ denote the ideal in $\mathcal{O}_U(V)$ generated by $\phi_1,\ldots,\phi_\ell$ and the determinants of the $(N-n)\times(N-n)$-minors in the jacobian matrix of $(F_1,\ldots,F_k,\phi_1,\ldots,\phi_\ell) : V \to \mathbb{C}^{k+\ell}$. It is not hard to check that $C(\phi_1,\ldots,\phi_\ell)$ is independent of the choice of generators: if $\psi_1,\ldots,\psi_m$ are $\mathcal{O}_U(V)$-linear combinations of $\phi_1,\ldots,\phi_\ell$: $\psi_\mu = \Sigma_\mu c_{\mu\lambda}\phi_\lambda$ then

$$\left(\frac{\partial\psi_\mu}{\partial z_\nu}\right) \equiv (c_{\mu\lambda}).\left(\frac{\partial\phi_\lambda}{\partial z_\nu}\right) \quad \mathrm{mod}(\phi_1,\ldots,\phi_\ell),$$

from which it easily follows that $C(\psi_1,\ldots,\psi_m) \subset C(\phi_1,\ldots,\phi_\ell)$. We let $C_f \subset \mathcal{O}_X$ $(= \mathcal{O}_U/\mathcal{I}_X)$ denote the corresponding ideal sheaf. In a similar fashion it follows that C_f only depends on f and not on the particular extension F. The following lemma says that C_f defines the critical locus of f.

(4.1) *Lemma*. Let $x \in X$ be such that $\dim(X,x) = n+k$. Then $C_{f,x} = \mathcal{O}_{X,x}$ if and only if X is nonsingular of dimension n+k at x and f is a submersion in x.

Proof. Let $\phi_1,\ldots,\phi_\ell$ generate $I_{X,x}$. If $C_{f,x} = O_{X,x}$, then there exists an (N-n)-tuple $F_{\kappa_1},\ldots,F_{\kappa_\alpha},\phi_{\lambda_1},\ldots,\phi_{\lambda_\beta}$ $(\alpha+\beta = N-n)$ taken out of $F_1,\ldots,F_k,\phi_1,\ldots,\phi_\ell$ whose differentials are linearly independent in x. Since $\dim(X,x) = n+k$, we must have $\beta \le N-(n+k)$. So after possibly renumbering the ϕ's, this (N-n)-tuple is $F_1,\ldots,F_k,\phi_1,\ldots,\phi_{N-n-k}$ and $\phi_1,\ldots,\phi_{N-n-k}$ generate $I_{X,x}$. The implicit function theorem then implies that we can find a local coordinate system $(z_1',\ldots,z_N')$ for $\mathbb{C}^N$ at x such that $F_\kappa = z'_{n+\kappa}$ $(\kappa = 1,\ldots,k)$ and $\phi_\lambda = z'_{n+k+\lambda}$ $(\lambda = 1,\ldots,N-n-k)$. The proof of the converse statement is left to the reader.

Quadratic singularities can also be nicely characterized in terms of C_f:

(4.2) *Proposition*. Let $f : (\mathbb{C}^m,0) \to (\mathbb{C}^k,0)$ be an analytic map-germ with $n := m-k \ge 0$. Then the following are equivalent

(i) $C_{f,0}$ is its own radical, $(C_f,0)$ is nonsingular of dim k-1 and $f|(C_f,0)$ is an immersion.

(ii) $C_{f,0} + (f_1,\ldots f_k)O_{\mathbb{C}^m,0} = m_{\mathbb{C}^m,0}$.

(iii) $(f_1,\ldots,f_k)$ defines a quadratic singularity of dim n.

(iv) There exist local coordinates for $(\mathbb{C}^m,0)$ and $(\mathbb{C}^k,0)$ in which f assumes the form $(z_1,\ldots,z_m) \to (z_1^2+\ldots+z_{m-k+1}^2, z_{m-k+2},\ldots,z_m)$.

Proof. (i) $\Rightarrow$ (ii). By the Nullstellensatz $C_{f,0}$ is the ideal of analytic functions vanishing on $(C_f,0)$. Since $f|(C_f,0)$ is an immersion it follows that $f_1,\ldots,f_k$ generate the maximal ideal of $O_{\mathbb{C}^m,0}/C_{f,0}$.

(ii) $\Rightarrow$ (iii). If r denotes the rank of df(0), then after coordinate changes in $(\mathbb{C}^m,0)$ and $(\mathbb{C}^k,0)$ we may assume that f has the form $f(z) = (f_1(z),\ldots,f_{k-r}(z),z_{m-r+1},\ldots,z_m)$ with $df_j(0) = 0$, $j = 1,\ldots,k-r$.

So $C_{f,0}$ will be generated by elements in $(m_{\mathbb{C}^m,0})^{k-r}$. Our assumption that $m_{\mathbb{C}^m,0}$ is generated by $C_{f,0}$ and $f_1,\ldots,f_k$ then implies that $k-r = 1$. So $f(z) = (f_1(z), z_{n+2},\ldots,z_m)$ with $df_1(0) = 0$ and $m_{\mathbb{C}^m,0}$ is generated by $\frac{\partial f_1}{\partial z_1},\ldots,\frac{\partial f_1}{\partial z_{n+1}}, f_1, z_{n+2},\ldots,z_m$. Since $df_1(0) = 0$, this implies that the $(n+1)\times(n+1)$-matrix $(\frac{\partial^2 f_1}{\partial z_\nu \partial z_\mu}(0) : 1 \le \nu,\mu \le n+1)$ is nonsingular. The complex-analytic form of the Morse Lemma (Milnor (1963), I.2.2), then implies that f defines a quadratic singularity of dim n.

(iii) ⇒ (iv). The assumption is that $\mathcal{O}_{\mathbb{C}^m,0}/(f_1,\ldots,f_k)\mathcal{O}_{\mathbb{C}^m,0}$ and $\mathcal{O}_{\mathbb{C}^{n+1},0}/(z_1^2+\ldots+z_{n+1}^2)\mathcal{O}_{\mathbb{C}^{n+1},0}$ are isomorphic $\mathbb{C}$-algebras. This implies that $df(0)$ is of rank $k-1$. Proceeding as above, we find that in possibly new coordinate systems, f has the form $f(z) = (f_1(z), z_{n+2},\ldots,z_m)$ with $f_1(z_1,\ldots,z_{n+1},0,\ldots,0) = (z_1^2+\ldots+z_{n+1}^2)\phi$, where ϕ is a unit of $\mathcal{O}_{\mathbb{C}^{n+1},0}$. It follows that $\frac{\partial f_1}{\partial z_1},\ldots,\frac{\partial f_1}{\partial z_{n+1}}, z_{n+2},\ldots,z_m$ have linearly independent differentials in 0. Since $C_{f,0}$ is generated by $\frac{\partial f_1}{\partial z_1},\ldots,\frac{\partial f_1}{\partial z_{n+1}}$, this implies that $(C_f,0)$ is a nonsingular transversal slice to the subspace of $\mathbb{C}^m$ defined by $z_{n+2} = \ldots = z_m = 0$. So by another coordinate change (involving $z_1,\ldots,z_{n+1}$ only) we may suppose that $\frac{\partial f_1}{\partial z_1},\ldots,\frac{\partial f_1}{\partial z_{n+1}}$ and $z_1,\ldots,z_{n+1}$ generate the same ideal in $\mathcal{O}_{\mathbb{C}^m,0}$. A trivial modification of the Morse lemma (loc. cit.) with $z_{n+2},\ldots,z_m$ as 'parameters' shows that after a suitable coordinate change (involving $z_1,\ldots,z_{n+1}$ only), f_1 assumes the form $z_1^2+\ldots+z_{n+1}^2$.

(iv) ⇒ (i) is easy.

(4.3) This proposition and the lemma preceding it suggest that the critical locus C_f should be endowed with the (coherent) structure sheaf $\mathcal{O}_X/\mathcal{C}_f$ (restricted to X). We shall do so: from now on C_f is always understood as being supplied with the sheaf $\mathcal{O}_{C_f} := \mathcal{O}_X/\mathcal{C}_f$. As this clashes with our tacit convention that $\mathcal{O}_{C_f}$ stands for $\mathcal{O}_X/\mathcal{I}_{C_f}$, we shall denote the latter by

$\mathcal{O}_{C_f,red}$. It is clear that this is obtained from $\mathcal{O}_{C_f}$ by killing the nilpotent elements of $\mathcal{O}_{C_f}$. If the two coincide in $x \in C_f$, we say that C_f is reduced in x. We shall refer to the ringed space $(C_f,\mathcal{O}_{C_f})$ as the *critical space* of f. For k = 0, we get the definition of the *singular space* of X, denoted by $(X_{sing},\mathcal{O}_{X_{sing}})$.

An even more important reason to define the critical space in this manner is that it 'commutes with base change'. This means that if we have a cartesian diagram of germs of analytic spaces

$$\begin{array}{ccc} (Y,y) & \xrightarrow{\tilde{h}} & (X,x) \\ {\scriptstyle g}\downarrow & & \downarrow{\scriptstyle f} \\ (\mathbb{C}^{\ell},0) & \xrightarrow{h} & (\mathbb{C}^k,0) \end{array}$$

(so $\mathcal{O}_{X,x}$ and $\mathcal{O}_{Y,y}$ may have nilpotent elements) with $\dim (f^{-1}(0),x) = n$, then $C_{g,y}$ is the ideal in $\mathcal{O}_{Y,y}$ generated by $\tilde{h}^*(C_{f,x})$. The proof is straightforward: if (X,x) is a germ in $\mathbb{C}^N$ at x defined by an ideal $(\phi_1,\ldots,\phi_r)\mathcal{O}_{\mathbb{C}^N,x}$, and $F_\kappa \in m_{\mathbb{C}^N,x}$ restricts to $f_\kappa \in m_{X,x}$ $(\kappa = 1,\ldots,k)$, then $C_{f,x}$ is the image in $\mathcal{O}_{X,x}$ of the ideal generated by the determinants of the $(N-n)\times(N-n)$-minors of the jacobian matrix of $(\phi_1,\ldots,\phi_r,F_1,\ldots,F_k)$: $(\mathbb{C}^N,x) \to \mathbb{C}^r\times\mathbb{C}^k$. The germ (Y,y) may be identified with the germ in $\mathbb{C}^N\times\mathbb{C}^\ell$ at (x,0) defined by the ideal I generated by $\pi_1^*(\phi_1),\ldots,\pi_1^*(\phi_r)$, $\pi_1^*(F_1)-\pi_2^*(h_1),\ldots,\pi_1^*(F_k)-\pi_2^*(h_k)$ (with $\pi_1 : \mathbb{C}^N\times\mathbb{C}^\ell \to \mathbb{C}^N$, $\pi_2 : \mathbb{C}^N\times\mathbb{C}^\ell \to \mathbb{C}^\ell$ denoting the projections) and g corresponds to π_2. This makes $C_{g,y}$ correspond to the image in $\mathcal{O}_{\mathbb{C}^N\times\mathbb{C}^\ell,x\times 0}/I$ of the ideal generated by the determinants of the $(N+\ell-n)\times(N\times\ell-n)$-minors of the jacobian matrix of $(\pi_1^*(\phi_1),\ldots,\pi_1^*(F_k)-\pi_2^*(h_k),s_1,\ldots,s_\ell) : (\mathbb{C}^N\times\mathbb{C}^\ell,x\times 0) \to \mathbb{C}^r\times\mathbb{C}^k\times\mathbb{C}^\ell$. Clearly this is the same ideal as the one generated by the determinants of the $(N-n)\times(N-n)$-minors of the jacobian matrix of $(\pi_1^*(\phi_1),\ldots,\pi_1^*(F_k)-\pi_2^*(h_k)) : (\mathbb{C}^N\times\mathbb{C}^\ell,x\times 0) \to \mathbb{C}^r\times\mathbb{C}^k$ with respect to the

$\mathbb{C}^N$-coordinates. The assertion follows. (The base change property is more elegantly expressed if we use the analytic tensor product $\hat{\otimes}$ as defined in Grauert & Remmert (1971) Ch. III, §5: the diagram being cartesian means that $\mathcal{O}_{\mathbb{C}^\ell,0} \hat{\otimes}_{\mathcal{O}_{\mathbb{C}^k,0}} \mathcal{O}_{X,x}$ maps isomorphically to $\mathcal{O}_{Y,y}$ and commutativity with base change says that this induces an isomorphism of $\mathcal{O}_{\mathbb{C}^\ell,0} \hat{\otimes}_{\mathcal{O}_{\mathbb{C}^k,0}} \mathcal{O}_{C_f,x}$ onto $\mathcal{O}_{C_g,y}$).

The particular case when h is the embedding of $\{0\}$ in $(\mathbb{C}^k,0)$ shows that $C_{f,x}+(f_1,\ldots,f_k)\mathcal{O}_{X,x}$ defines $(f^{-1}(0)_{sing},x)$.

Example 1. If $f : (\mathbb{C}^{n+1},x) \to (\mathbb{C},0)$ is an analytic germ, then

$$\mathcal{O}_{C_f,x} = \mathcal{O}_{\mathbb{C}^{n+1},x} / (\frac{\partial f}{\partial z_1},\ldots,\frac{\partial f}{\partial z_{n+1}})\mathcal{O}_{\mathbb{C}^{n+1},x},$$

$$\mathcal{O}_{f^{-1}(0)_{sing},x} = \mathcal{O}_{\mathbb{C}^{n+1},x} / (f,\frac{\partial f}{\partial z_1},\ldots,\frac{\partial f}{\partial z_{n+1}})\mathcal{O}_{\mathbb{C}^{n+1},x}.$$

We next look at the complete intersection case. A basic result is:

(4.4) *Proposition.* Let $f : (\mathbb{C}^{n+k},x) \to (\mathbb{C}^k,0)$ define an icis of dim n. Then $\mathcal{O}_{C_f,x} = \mathcal{O}_{\mathbb{C}^{n+k},x} / C_{f,x}$ is a Cohen-Macaulay ring. Also, $\mathcal{O}_{C_f,x}$ is a Cohen-Macaulay $\mathcal{O}_{\mathbb{C}^k,0}$-module of finite type.

Proof. If $(C,x) := (C_f,x)$ is non-vacuous, then (2.8.iii) says that its dim is k-1. Since (C,x) is defined by $C := C_{f,x}$, this amounts to saying that the C-depth of $\mathcal{O}_{\mathbb{C}^{n+k},x}$ (the length of a maximal regular sequence in C) equals n+1. According to Buchsbaum & Rim (1964), Cor. (2.7), this, in turn, implies that the homological dimension of $\mathcal{O}_{C,x}$ as an $\mathcal{O}_{\mathbb{C}^{n+k},x}$-module also equals n+1. But $\mathcal{O}_{\mathbb{C}^{n+k},x}$ is a regular ring and hence the latter is equivalent to $\mathcal{O}_{C,x}$ having depth k-1, (see for instance Serre (1975), Ch. IV, Prop. 21). As $k-1 = \dim(C,x)$ it follows that $\mathcal{O}_{C,x}$ is a Cohen-Macaulay ring.

According to Malgrange's version of the Weierstrass preparation theorem (Narasimhan (1966), Ch.II, Th.1), $\mathcal{O}_{C,x}$ is an $\mathcal{O}_{\mathbb{C}^k,0}$-module of finite type if (and only if) $\mathcal{O}_{C,x}/m_{\mathbb{C}^k,0}\mathcal{O}_{C,x}$ ($\cong \mathcal{O}_{\mathbb{C}^{n+k},x}/(C+(f_1,\dots,f_k)\mathcal{O}_{\mathbb{C}^{n+k},x})$) has finite $\mathbb{C}$-dimension. This condition is fulfilled: as $f^{-1}(0)$ has an isolated singular point at x, defined by the ideal $C+(f_1,\dots,f_k)\mathcal{O}_{\mathbb{C}^{n+k},x}$, this ideal must contain a power of $m_{\mathbb{C}^{n+k},0}$ (Nullstellensatz) and hence be of finite $\mathbb{C}$-codimension. Since $\mathcal{O}_{C,x}$ is a Cohen-Macaulay ring, it now follows that $\mathcal{O}_{C,x}$ is also Cohen-Macaulay, when viewed as an $\mathcal{O}_{\mathbb{C}^k,0}$-module.

An important property of a Cohen-Macaulay R-module M (with R a local noetherian ring) is its equidimensionality: if p is an associated prime of M (a prime ideal of R which is also the annihilator of some $m \in M$), then dim R/p is independent of p, hence equal to dim M (Serre (1975), Ch.IV, Prop.13). So if X is an analytic space and $x \in X$ is such that $\mathcal{O}_{X,x}$ is Cohen-Macaulay, then all the irreducible components of the germ (X,x) have the same dimension (n, say). It also follows, that there is a neighbourhood U of x in X such that the set of $y \in U$ with $\mathcal{O}_{X,y}$ nonreduced is a union of irreducible components of U. To see this, we suppose $\mathcal{O}_{X,x}$ nonreduced (otherwise there is a U such that $\mathcal{O}_{U,y}$ is reduced for all $y \in U$ by the Oka coherence theorems) and choose a nonzero nilpotent $\phi \in \mathcal{O}_{X,x}$. Let U be a neighbourhood of x in X such that ϕ is a section of $\mathcal{O}_U$ and put $Y = \mathrm{supp}(\phi)$. Each irreducible component (Y_i,x) of (Y,x) is defined by a prime ideal $p_i \subset \mathcal{O}_{X,x}$ which contains $\mathrm{Ann}(\phi)$ and is minimal for this property. Such a prime is necessarily associated to $\mathrm{Ann}(\phi)$ (Serre (1975), Ch.I, Th.1). In particular, p_i is associated to $\mathcal{O}_{X,x}$ and so $\dim(Y_i,x) = \dim(\mathcal{O}_{X,x}/p_i) = n$. Hence (Y,x) is a union of irreducible components of (X,x).

We use this to interpret the condition that the critical space be reduced.

(4.5) *Proposition.* Let $f : (\mathbb{C}^{n+k},x) \to (\mathbb{C}^k,0)$ define an icis of dim n. Then $\mathcal{O}_{C_f,x}$ is reduced if and only if the generic singularity of f along (C_f,x) is quadratic (by which we mean that f admits a representative $f : X \to \mathbb{C}^k$ such that the set of $y \in C_f$ where f exhibits a quadratic singularity is dense in C_f).

Proof. If $\mathcal{O}_{C,x}$ is reduced, then choose a good representative $f : X \to S$ of f whose critical space C is reduced. By (2.8) ii,iii C is of pure dimension k-1 and $f|C$ is finite. So the set of $z \in C_f$ where C is singular or $f|C$ is not an immersion is an analytic subset C' of C of dim < k-1. In particular, it is nowhere dense in C. Following (4.2) (i) $\Leftrightarrow$ (iii), C' is just the set of $y \in C$ where f fails to exhibit a quadratic singularity. Conversely, suppose that there is a representative $f : X \to \mathbb{C}^k$ such that the set of $y \in C_f$ where f has a quadratic singularity is dense in C_f. Since $\mathcal{O}_{C_f,y}$ is reduced at such points and $\mathcal{O}_{C_f,x}$ is Cohen-Macaulay, the preceding discussion shows that $\mathcal{O}_{C_f,x}$ is also reduced.

4.B *The Thom singularity manifolds*

Let m and k be nonnegative integers. We define for $r \leq \min\{m,k\}$ the *Thom stratum*

$$\Sigma_r(m,k) := \{\sigma \in \mathrm{Hom}(\mathbb{C}^m,\mathbb{C}^k) : \mathrm{rank}(\sigma) = r\}.$$

The resulting partition of $\mathrm{Hom}(\mathbb{C}^m,\mathbb{C}^k)$ (by rank) is called the *Thom stratification* of $\mathrm{Hom}(\mathbb{C}^m,\mathbb{C}^k)$. If $\sigma \in \Sigma_r(m,k)$, then it is always possible to find bases $e_1,\ldots,e_m$ for $\mathbb{C}^m$ and $f_1,\ldots,f_k$ for $\mathbb{C}^k$ with respect to which σ

takes the standard form: $\sigma(e_i) = f_i$ $(i = 1,\ldots,r)$, $\sigma(e_i) = 0$ $(i > r)$. This implies that $\Sigma_r(m,k)$ is a single $GL_m(\mathbb{C})\times GL_k(\mathbb{C})$-orbit under the natural representation of the latter on $Hom(\mathbb{C}^m,\mathbb{C}^k)$. Using this for $\Sigma_{r'}(m,k)$ $(r' < r)$ we also find that

$$\overline{\Sigma}_r(m,k) = \Sigma_r(m,k) \cup \Sigma_{r-1}(m,k) \cup \ldots \cup \Sigma_0(m,k).$$

Now, $\overline{\Sigma}_r(m,k)$ being the common zero set of the determinants of the $(r+1)\times(r+1)$-minors, is an affine algebraic variety, so that $\Sigma_r(m,k)$ is Zariski-constructible. We claim that $\Sigma_r(m,k)$ is nonsingular of codim $(m-r)(k-r)$ in $Hom(\mathbb{C}^m,\mathbb{C}^k)$. We need only verify this at the standard representative

$$\sigma_0 = \begin{pmatrix} 1_r & 0 \\ 0 & 0 \end{pmatrix}$$

Let $U \subset Hom(\mathbb{C}^m,\mathbb{C}^k)$ denote the Zariski-open set of

$$\tau = \begin{pmatrix} \tau_{11} & \tau_{12} \\ \tau_{21} & \tau_{22} \end{pmatrix}$$

with τ_{11} invertible and define $\pi : U \to Hom(\mathbb{C}^{m-r},\mathbb{C}^{k-r})$ by $\pi(\tau) = \tau_{22} - \tau_{21}\tau_{11}^{-1}\tau_{12}$. It is clear that π is a submersion everywhere. Any $\tau \in U$ can be written

$$\begin{pmatrix} \tau_{11} & 0 \\ \tau_{21} & 1_{k-r} \end{pmatrix}\begin{pmatrix} 1_r & \tau_{11}^{-1}\tau_{12} \\ 0 & \pi(\tau) \end{pmatrix}$$

in which the first $k\times k$-matrix is nonsingular. So $\Sigma_r(m,k) \cap U$ is just the fibre $\pi^{-1}(0)$. This certainly implies our claim. Our argument also shows that an analytic map-germ $F : (\mathbb{C}^\ell,x) \to (Hom(\mathbb{C}^m,\mathbb{C}^k),\sigma_0)$ is transversal to $\Sigma_r(m,k)$ if and only if $\pi\circ F$ is a submersion. This, in turn, is equivalent to F_{22} being a submersion. We then say that F is *Thom-transversal*. Since $\Sigma_{r'}(m,k) \cap U = \pi^{-1}(\Sigma_{r'-r}(m-r,k-r))$, there is a representative of F defined on some open $V \ni x$ which is Thom-transversal at every point of V.

So Thom-transversality is an open property. Also, if we are given an analytic $g : (\mathbb{C}^\ell,x) \to GL_m(\mathbb{C})\times GL_k(\mathbb{C})$, then F is Thom-transversal if and only if $z \to g(z).F(z)$ is (this is merely a consequence of the $GL_m(\mathbb{C})\times GL_k(\mathbb{C})$-invariance of the stratification).

If $f : (\mathbb{C}^m,x) \to \mathbb{C}^k$ is an analytic map-germ, then by abuse of language 'f is Thom-transversal' simply means that $df : (\mathbb{C}^m,x) \to \mathrm{Hom}(\mathbb{C}^m,\mathbb{C}^k)$ is. So in that case f has a representative defined on some open $V \ni x$ such that

$$\Sigma_r(f) := \{z \in V : \mathrm{rank}\ df(z) = r\}$$

is an analytically constructible submanifold of V of codim $(m-r)(k-r)$. It also follows from the above discussion that if $df(z)$ is of rank r, then the partition of the $\Sigma_{r'}(f)$ near x is analytically equivalent to the product of the Thom stratification of $\mathrm{Hom}(\mathbb{C}^{m-r},\mathbb{C}^{k-r})$ at the origin and a nonsingular germ (of dim $m-(m-r)(k-r)$). In particular, $\overline{\Sigma}_{r'}(f) = \Sigma_{r'}(f) \cup\ldots\cup \Sigma_0(f)$ and $\overline{\Sigma}_{r'}(f)$ is locally irreducible.

Finally notice that the property of $f : (\mathbb{C}^m,x) \to \mathbb{C}^k$ being Thom-transversal is coordinate-invariant.

(4.6) *Lemma*. Any of the conditions of (4.2) is equivalent to

(v) $f : (\mathbb{C}^{n+k},0) \to (\mathbb{C}^k,0)$ is Thom-transversal, $df(0)$ has rank $k-1$ and $f|C_f$ is an immersion.

Proof. (v) ⇒ (i). If $df(0)$ has rank $k-1$, then we may assume that $df_1(0) = 0$ and $f_\kappa(z) = z_{n+\kappa}$ $(\kappa = 2,\ldots,k)$. Thom-transversality is then equivalent to $\left(\frac{\partial f_1}{\partial z_1},\ldots,\frac{\partial f_1}{\partial z_{n+1}}\right) : \mathbb{C}^{n+k},0 \to \mathbb{C}^{n+1},0$ being a submersion. Since C_f is the ideal generated by these partial derivatives, (v) ⇒ (i) follows.

(iv) ⇒ (v) is easy.

4.C *Development of the discriminant locus*

We first describe a resolution for the closed Thom stratum $\overline{\Sigma}_{k-1}(n+k,k)$ of singular $(n+k)\times k$ matrices. Put

$$\widetilde{\Sigma}_{k-1}(n+k,k) = \{(\sigma,\alpha) \in \mathrm{Hom}(\mathbb{C}^{n+k},\mathbb{C}^k)\times\check{\mathbb{P}}^{k-1} : \mathrm{Image}\ (\sigma)\subset\alpha\}.$$

Here $\check{\mathbb{P}}^{k-1}$ denotes the projective space of hyperplanes in $\mathbb{C}^k$. Notice that $\widetilde{\Sigma}_{k-1}(n+k,k)$ is just the total space of $(\gamma^{k-1})^{\oplus(n+k)}$, where γ^{k-1} is the tautological (hyperplane) bundle over $\check{\mathbb{P}}^{k-1}$. In particular, $\widetilde{\Sigma}_{k-1}(n+k,k)$ is nonsingular. The projection π onto the first factor maps $\widetilde{\Sigma}_{k-1}(n+k,k)$ onto $\overline{\Sigma}_{k-1}(n+k,k)$ and this map is an isomorphism over $\Sigma_{k-1}(n+k,k)$. So this is indeed a resolution of $\Sigma_{k-1}(n+k,k)$. The exceptional locus is contained in (actually equal to) the pre-image of $\overline{\Sigma}_{k-2}(n+k,k)$ and a simple calculation shows that its codimension therefore is $\geq n+2$.

Now let $f : (\mathbb{C}^{n+k},0)\to(\mathbb{C}^k,0)$ be a Thom-transversal map-germ which defines an icis. Choose a good representative $f : X\to S$ which is Thom-transversal. Then we define $\widetilde{C}$ by the cartesian diagram

$$\begin{array}{ccc} \widetilde{C} & \longrightarrow & \widetilde{\Sigma}_{k-1}(n+k,k) \\ \pi_f \downarrow & & \downarrow \pi \\ X & \xrightarrow{df} & \mathrm{Hom}(\mathbb{C}^{n+k},\mathbb{C}^k) \end{array}$$

Since f is Thom-transversal, df and π are mutually transverse so that $\widetilde{C}$ is nonsingular and π_f defines a resolution of the critical locus C. In invariant terms, $\widetilde{C}$ is simply $\{(z,\alpha)\in X\times_S\check{\mathbb{P}}(\tau_S) : \mathrm{Image}\ df(z)\subset\alpha\}$, where τ_S denotes the tangent bundle of S and $\check{\mathbb{P}}(\tau_S)$ the Grassmann bundle of its hyperplanes. A remarkable aspect of this situation is that both C and $\widetilde{C}$ can be reconstructed from the discriminant D of f. Recall from (2.8) that in this case D is a hypersurface in S. If D_{reg} stands for its set of regular points, let $\widetilde{D}_{reg}$ denote the image of the section $s \in D_{reg} \to T_s(D_{reg}) \in \check{P}(\tau_S)$. Then the closure $\widetilde{D}$ of $\widetilde{D}_{reg}$ in $\check{P}(\tau_S)$ is what Teissier calls the *development* of D. This is an analytic subset of $\check{P}(\tau_S) \cong S\times\check{P}^{k-1}$: if δ generates I_D over an open $U \subset S$, then $\widetilde{D} \cap (U\times\check{P}^{k-1})$ is the union of the irreducible components of the analytic variety in $U\times\check{P}^{k-1}$ defined by

$$\{(s,[\lambda_1:\ldots:\lambda_k]) \in U \times \check{P}^{k-1}: \delta(s) = 0,$$

$$\lambda_i \frac{\partial\delta}{\partial t_j} - \lambda_j \frac{\partial\delta}{\partial t_i} = 0,\ 1 \le i,j \le k\}$$

which meet $\widetilde{D}_{reg}$. If $z \in \Sigma_{k-1}(f)$ is such that C is smooth in z, $f|C$ is an immersion in z and $f(z) \in D_{reg}$, then f defines a quadratic singularity at z by (4.2) i. If we put $\alpha := df(T_zX)$, then it follows from the local model (4.2) iv, that $\alpha = T_{f(z)}(D_{reg})$. So (z,α) is the unique point of $\widetilde{C}$ lying over z and $(f(z),\alpha)$ is the unique point of $\widetilde{D}$ lying over f(z). Since such z resp. f(z) are dense in C resp. D it follows that we have a surjective map $\widetilde{f} : \widetilde{C} \to \widetilde{D}$. The following theorem is due to Teissier (1976, Thm. 5.5.1) in case f defines a hypersurface singularity (and satisfies a somewhat stronger condition than Thom-transversality).

(4.7) *Theorem*. In the situation above, we have after possible shrinking of S that C is reduced and normal, $f : C \to D$ is a normalization of D, D is irreducible and $\widetilde{f} : \widetilde{C} \to \widetilde{D}$ is an isomorphism.

The geometric implications of this theorem are quite interesting. For instance, the fact that $f : C \to D$ is a normalization implies that this map is of degree one, which is not at all obvious. The fact that $\widetilde{f}$ is an isomorphism shows among other things the following: if $\{\widetilde{s}_i \in \widetilde{D}_{reg}\}_{i=1}^{\infty}$ tends to $0 \in S$ and the sequence $T_{s_i}(D_{reg})$ tends to a hyperplane α in $T_0(S)$, then α contains the image of $T_0(X)$ under $df(0)$. Moreover any hyperplane containing the image of $df(0)$ arises in this way. It also follows that if z_1 and z_2 are distinct points of C lying over the same point $s \in D$, then there is no hyperplane in $T_s(D)$ containing the images of both $T_{z_1}(X)$ and $T_{z_2}(X)$, in other words $df : T_{z_1}(X) \to T_s(D)$ and $df : T_{z_2}(X) \to T_s(D)$ are mutually transverse. Another surprising consequence is that $\tilde{D}$ is nonsingular.

Finally we notice that if $df(0)$ has rank $k-1$ (which just means that $(X_0,0)$ is a hypersurface germ), then $\tilde{C}$ is nonsingular at 0 and $\pi_f : \tilde{C} \to C$ is an isomorphism over 0. So in this case the normalization and the development of D coincide (and are nonsingular) over 0.

Proof of (4.7). Since the quadratic singularities are dense in C (we saw this in the discussion preceding the theorem), it follows from (4.5) that C is reduced. Recall also that $\mathcal{O}_C$ is Cohen-Macaulay, so normality of C will follow if we show that C is nonsingular in codim 1 (Serre (1975), Ch. IV, Th. 11 or Gunning (1974), Cor. 1 to Th. 15). This is certainly the case, as the singular locus of C is contained in $\overline{\Sigma}_{k-2}(f)$ which is of codim n+3 in C. The remainder will be proved in four steps.

Step 1. If $\psi \in m_{\mathbb{C}^k,0}$ is such that $d\psi(0) \neq 0$, then

$$m_{\mathbb{C}^{n+k},0} \subset (f_1,\ldots,f_k,\frac{\partial}{\partial z_1}(\psi\circ f),\ldots,\frac{\partial}{\partial z_{n+k}}(\psi\circ f))\mathcal{O}_{\mathbb{C}^{n+k},0}.$$

If $\frac{\partial}{\partial z_\nu}(\psi\circ f)(0) \neq 0$ for some ν, then the right hand side clearly equals $\mathcal{O}_{\mathbb{C}^{n+k},0}$ and so we are done. So let us suppose that $d(\psi\circ f)(0) = 0$. Our strategy is to choose now coordinates in $(\mathbb{C}^k,0)$ and $(\mathbb{C}^{n+k},0)$ with respect to which the assertion of step 1 is fairly obvious. For this it is important to note that this assertion is indeed coordinate invariant.

Let r denote the rank of $df(0)$. Then we can choose $v_1,\ldots,v_r \in m_{S,0}$ such that $u_1 := v_1\circ f,\ldots,\ u_r := v_r\circ f$ have linearly independent differentials in 0. Since $d\psi(0) \neq 0$ and $d(\psi\circ f) = 0$, we can extend $v_1,\ldots,v_r$ to a coordinate system $v_1,\ldots,v_k$ for $(\mathbb{C}^k,0)$ such that $\psi = v_{k+1}$. If $u_1,\ldots,u_r$ is extended to a coordinate system $(u_1,\ldots,u_r,x_1,x_2,\ldots)$ for $(\mathbb{C}^{n+k},0)$, then the quadratic part of $\psi\circ f$ will be of the form

$$\tilde{q}(x) + \ell_1(u)x_1 + \ell_2(u)x_2 + \ldots + q(u)$$

where $\tilde{q}$ and q are quadratic forms and $\ell_1,\ell_2,\ldots$ are linear forms. By choosing $x_1,x_2,\ldots$ carefully, we may assume that $\tilde{q}$ has the diagonal form $x_1^2 + \ldots + x_s^2$ (for some s). By splitting off squares (replace x_σ by $x_\sigma + \frac{1}{2}\ell_\sigma$) we may moreover assume that $\ell_1 = \ldots = \ell_s = 0$, so that we end up with

$$\psi\circ f = x_1^2 + \ldots + x_s^2 + \ell_{s+1}(u)x_{s+1} + \ldots + \ell_{s+t}(u)x_{s+t} + q(u) + \text{h.o.t.}$$

where $t = (n+k)-(r+s)$. Since the derivative of f in 0 is already in standard form with respect to these coordinates, the Thom-transversality of f implies that $\frac{\partial}{\partial x_1}(\psi\circ f),\ldots,\frac{\partial}{\partial x_{s+t}}(\psi\circ f)$ have linearly independent differentials in 0. This just means that $\ell_{s+1},\ldots,\ell_{s+t}$ are linearly independent. By an appropriate linear transformation in the $(v_1,\ldots,v_r)$-coordinate space we may therefore even assume that $\ell_{s+j}(u) = u_j$. It is then clear that up to higher order terms each of the coordinates $x_1,\ldots,x_{s+t}$ occurs as a partial derivative of $\psi\circ f$. Since $u_\rho = v_\rho\circ f$, the assertion now follows from Nakayama's lemma.

Step 2. If $\delta \in I_D$ then $\frac{\partial}{\partial z_\nu}(\delta\circ f) \in \mathcal{C}$ for all ν.

Since C is its own radical it suffices to verify this at a generic point x of C. If f has x as a quadratic singularity then this is immediate from the local model (4.2)-iv (in the notation used there, $\delta\circ f$

is divisible by f_1 and hence vanishes of order two on C).

Step 3. $\tilde{f} : \tilde{C} \to \tilde{D}$ is an isomorphism over a neighbourhood V of 0 in S.

Let $\tilde{\lambda} = (0,\lambda) \in \tilde{C} \cap (\{0\} \times \mathbf{P}^{k-1})$. We first show that $\tilde{f}$ is an immersion in $\tilde{\lambda}$. Without loss of generality we assume that $\lambda = [0:\ldots:0:1]$. Then we have affine coordinates $\ell_1,\ldots,\ell_{k-1}$ for $\mathbf{P}^{k-1}$ at λ given by $[\ell_1:\ldots:\ell_{k-1}:1]$. Now let δ generate $I_{D,0}$. Then

$$\frac{\partial}{\partial z_\nu}(\delta \circ f) = \Sigma_{j=1}^{k} f^*(\frac{\partial \delta}{\partial t_j})\frac{\partial f_j}{\partial z_\nu}$$

and by step 2 the left hand side is in C_0. In $\tilde{\lambda}$ we have

$$\tilde{f}^*(\frac{\partial \delta}{\partial t_j}) = \ell_j \tilde{f}^*(\frac{\partial \delta}{\partial t_k}) \quad (j=1,\ldots,k-1)$$

It follows that in $\mathcal{O}_{\tilde{C},\tilde{\lambda}}$ we have

$$(*) \qquad -\pi_f^*(\frac{\partial f_k}{\partial z_\nu}) = \Sigma_{j=1}^{k-1} \ell_j \pi_f^*(\frac{\partial f_j}{\partial z_\nu}).$$

Hence

$$(z_1,\ldots,z_{n+k})\mathcal{O}_{\tilde{C},\tilde{\lambda}} \subset (f_1,\ldots,f_k,\frac{\partial f_k}{\partial z_1},\ldots,\frac{\partial f_k}{\partial z_{n+k}})\mathcal{O}_{\tilde{C},\tilde{\lambda}} \text{ (by step 1)}$$

$$\subset (f_1,\ldots,f_k,\ell_1,\ldots,\ell_{j-1})\mathcal{O}_{\tilde{C},\tilde{\lambda}} \quad \text{(by (*))}$$

$$= m_{\tilde{D},\tilde{f}(\tilde{\lambda})}\mathcal{O}_{\tilde{C},\tilde{\lambda}}$$

from which it follows that $m_{\tilde{C},\tilde{\lambda}}\mathcal{O}_{\tilde{C},\tilde{\lambda}} = m_{\tilde{D},\tilde{f}(\tilde{\lambda})}\mathcal{O}_{\tilde{C},\tilde{\lambda}}$. This proves that $\tilde{f}$ is an immersion in $\tilde{\lambda}$. Since $\tilde{f}$ restricted to $\tilde{C}\times(\{0\}\times\mathbf{P}^{k-1})$ is an injection, it follows that $\tilde{f}$ is an embedding over a neighbourhood of $0 \in X$. Then $\tilde{f}$ is also an embedding over a neighbourhood V of $0 \in S$, because $f^{-1}(0) \cap C \subset \{0\}$. The assertion follows from this and the fact that $\tilde{f}$ is surjective.

Step 4. $f : C \to D$ is a normalization over V and $D \cap V$ is irreducible.

Since C is normal and $f : C \to D$ is finite we only have to show that $f : C \to D$ is of degree one over V. This is clear, since the fact that $\tilde{f}$ is an isomorphism over V implies that any point of $\tilde{D}_{reg} \cap V$ has exactly one pre-image in C. Since $C \cap f^{-1}(V)$ is connected, $D \cap V$ is irreducible.

4.D *Fitting ideals*

This is going to be a short excursion into commutative algebra, necessary to give a proper definition of the discriminant space.

We start out with a commutative ring R with 1 and a finitely generated (unitary) R-module M. So there is a surjective homomorphism $\alpha : R^p \to M$. Let us denote the natural basis vectors of R^p by $e_1,\dots,e_p$. For any $(p-k)$-tuple $x_1,\dots,x_{p-k}$, $x_i = \Sigma_{j=1}^{p}\xi_{ij}e_j$, taken from $\text{Ker}(\alpha)$ and any increasing sequence $1 \le j_1 <\dots< j_{p-k} \le p$ we can form the determinant of the $(p-k)\times(p-k)$ matrix (ξ_{ij_ℓ}). Let us provisionally denote by $F_k(\alpha)$ the ideal in R generated by such determinants (we put $F_k(\alpha) = R$ if $k \ge p$). It suffices to let $(x_1,\dots,x_{p-k})$ run over the $(p-k)$-tuples in a generating set of $\text{Ker}(\alpha)$. In particular, if $\text{Ker}(\alpha)$ is finitely generated, then so is $F_k(\alpha)$. We show that $F_k(\alpha)$ only depends on M. If $\beta : R^q \to M$ is another surjective homomorphism and the basis vectors of R^q are denoted $f_1,\dots,f_q$, then for any f_i we choose $y_i \in R^p$ such that $\alpha(y_i) = -\beta(f_i)$. Then it is easy to see that the kernel of $\alpha+\beta : R^p \oplus R^q \to M$ is generated by $(\text{Ker}(\alpha),0)$ and the pairs $\{(y_i,f_i) : i=1,\dots,q\}$. If we pick our $p+q-k$ elements in this generating set then we find that $F_k(\alpha+\beta) = F_k(\alpha)$. In the same way we have $F_k(\alpha+\beta) = F_k(\beta)$.

So $F_k(\alpha)$ is an invariant of M. It is called the kth *Fitting ideal* and denoted by $F_k(M)$. Since the determinant of a k×k matrix is a linear combination of the determinants of its (k-1)×(k-1) minors, we have an ascending chain of ideals

$$F_0(M) \subset F_1(M) \subset F_2(M) \subset \ldots$$

If $\sigma : R^p \to R^p$ is a homomorphism, then Cramer's rule shows that $\det(\sigma).R^p$ is contained in the image of σ. It follows that $F_0(M)$ is always contained in the annihilator Ann(M) of M. In particular, M has the structure of a $R/F_0(M)$-module.

Example 2. If V is a finite dimensional vector space over a field K, then $F_j(V) = 0$ if $j < \dim V$ and $F_j(V) = K$ if $j \geq \dim V$.

Example 3. Let R be a principal ideal domain. Then any finitely generated R-module M is isomorphic to

$$R/I_1 \oplus R/I_2 \oplus \ldots \oplus R/I_r$$

where $I_1 \supset I_2 \supset \ldots \supset I_r$ is a descending sequence of (principal) ideals $\neq R$, so $F_k(M) = I_1 I_2 \ldots I_{r-k}$ if $k < r$ and $F_k(M) = R$ otherwise. Note that $\mathrm{Ann}(M) = I_1$ is in general different from $F_0(M) = I_1 \ldots I_r$. If none of the ideals I_ρ is trivial, then M has finite length: if π_ρ is the number of prime factors in a generator of I_ρ, then $\ell(M) = \pi_1 + \ldots + \pi_r$. This is also the length of $R/(I_1 \ldots I_r)$.

An important property of Fitting ideals is that their formation commutes with base change: if $\alpha : R^p \to M$ is a surjective R-homomorphism and $\phi : R \to S$ is a homomorphism of rings with $\phi(1) = 1$, then tensoring with S gives a surjective S-homomorphism $S^p \to S \otimes_R M$. The kernel of the latter is generated by $\phi^p(\mathrm{Ker}(\alpha))$ and so $F_k(S \otimes_R M)$ is the ideal in S generated by $\phi(F_k(M))$.

The notion of a Fitting ideal also makes sense if we are given an analytic space $(X,\mathcal{O}_X)$ and a coherent $\mathcal{O}_X$-module M. Then the kth *Fitting ideal sheaf* $F_k(M)$ is simply the sheaf associated to the presheaf which assigns to an open $U \subset X$ the kth Fitting ideal of the $\Gamma(U,\mathcal{O}_X)$-module $\Gamma(U,M)$. The fact that the formation of the kth Fitting ideal commutes with base change is now expressed by the property that if $g : (Y,\mathcal{O}_Y) \to (X,\mathcal{O}_X)$ is a morphism of analytic spaces, the kth Fitting ideal of the pull-back $g^*(M)$ $(:= \mathcal{O}_Y \otimes_{\mathcal{O}_X} M)$ of M is equal to $g^*(F_k(M))$. If we take for Y just a single point $x \in X$, we find that $F_k(M_x)/m_{X,x}F_k(M_x) = F_k(M_x/m_{X,x}M_x)$. According to example 2 the latter is zero if and only if $\dim_{\mathbb{C}} M_x/m_{X,x}M_x > k$. So $F_k(M)$ defines the locus of $x \in X$, over which the 'fibre' $M_x/m_{X,x}M_x$ of M has dim $> k$. In particular, $F_0(M)$ defines the support of M.

4.E *The discriminant space*

Following Rim (1972) and Teissier (1972) we shall define the image of a finite morphism $f : (Y,\mathcal{O}_Y) \to (S,\mathcal{O}_S)$ of analytic spaces *as an analytic space* by means of Fitting ideals. Recall that 'finite' means proper and with finite fibres. The direct image $f_*\mathcal{O}_Y$ is then a coherent sheaf of $\mathcal{O}_S$-modules (Narasimhan (1966), Ch. IV, Th. 7), whose support is of course $f(Y)$. This is also the support of $\mathcal{O}_S/F_0(f_*\mathcal{O}_Y)$. We endow $f(Y)$ with the structure sheaf $\mathcal{O}_S/F_0(f_*(\mathcal{O}_Y))$ (restricted to $f(Y)$) and denote the analytic space thus defined by $(f(Y), \mathcal{O}_{f(Y)})$. What makes this the right way to define $\mathcal{O}_{f(Y)}$? The answer is that it is the only conceivable definition which behaves well under base change: if

$$\begin{array}{ccc} (Z,\mathcal{O}_Z) & \xrightarrow{\tilde{g}} & (Y,\mathcal{O}_Y) \\ \tilde{f}\downarrow & \square & \downarrow f \\ (T,\mathcal{O}_T) & \xrightarrow{g} & (S,\mathcal{O}_S) \end{array}$$

is a cartesian diagram of analytic spaces with f (and hence $\tilde{f}$) finite, then $\tilde{f}(Z) = g^{-1}(f(Y))$. But we also have $\tilde{f}_*\mathcal{O}_Z \cong \mathcal{O}_T \otimes_{\mathcal{O}_S} f^*\mathcal{O}_Y$ (for f is finite). Hence $F_0(\tilde{f}_*\mathcal{O}_Z)$ is generated by $g^*F_0(f_*\mathcal{O}_Y)$, which is equivalent to saying that $\mathcal{O}_{\tilde{f}(Z)} \cong \mathcal{O}_T \otimes_{\mathcal{O}_S} \mathcal{O}_{f(Y)}$.

We further notice that since $F_0(f_*\mathcal{O}_Y)$ is contained in the annihilator of $f_*\mathcal{O}_Y$, $f_*\mathcal{O}_Y$ acquires the structure of an $\mathcal{O}_{f(Y)}$-module.

The case we have in mind of course, is of the discriminant of a good representative $f : X \to S$, which appears as the image of the critical locus C under the finite map $f : C \to S$. So we let the *discriminant space* $(D,\mathcal{O}_D)$ of f be the f-image in the above sense of the critical space $(C,\mathcal{O}_C)$. This notion commutes with base change since the formation of critical space and image do.

Example 4. Let $f : (\mathbb{C}^{n+1},x) \to (\mathbb{C},0)$ define an isolated hypersurface singularity. Then $C = \{x\}$, $D = \{0\}$ and $\mathcal{O}_{C,x} = \mathcal{O}_{\mathbb{C}^{n+1},x}/(\frac{\partial f}{\partial z_1},\ldots,\frac{\partial f}{\partial z_{n+1}})\mathcal{O}_{\mathbb{C}^{n+1},x}$. The latter is by (1.2) a finite-dimensional $\mathbb{C}$-vector space whose dimension we denote by μ. So $\mathcal{O}_{C,x}$ viewed as an $\mathcal{O}_{\mathbb{C},0}$-module has length μ. Now $\mathcal{O}_{\mathbb{C},0} = \mathbb{C}\{t\}$ is a principal ideal domain so that by example 3, $\mathcal{O}_{D,0} = \mathcal{O}_{\mathbb{C},0}/F_0(\mathcal{O}_{C,x})$ also has length μ. It follows that $\mathcal{O}_{D,0} = \mathbb{C}\{t\}/t^{\mu}\mathbb{C}\{t\}$.

Example 5. More generally, let $f : (\mathbb{C}^{n+k},x) \to (\mathbb{C}^k,0)$ define an icis of dim n and assume that the last coordinate axis ℓ of $\mathbb{C}^k$ meets the discriminant $(D,0)$ of f in the origin only. Put $(X_0',x) = (f^{-1}(\ell),x)$ and let $\tilde{f}$ $(= f_k) : (Y,x) \to (\mathbb{C},0)$ denote the restriction. The defining ideal of D_f induces in $\mathbb{C}\{t_k\}$ the defining ideal of $D_{\tilde{f}}$ (commutativity with base change) and the argument used in example 4 shows that this is $t_k^\nu \mathbb{C}\{t_k\}$, where ν is the $\mathbb{C}$-dimension of

$$\mathcal{O}_{\mathbb{C}^{n+k},x} / (\{\frac{\partial(f_1,\ldots,f_k)}{\partial(z_{\nu_1},\ldots,z_{\nu_k})} : 1 \leq \nu_1 <\ldots< \nu_k \leq n+k\},$$
$$f_1,\ldots,f_{k-1})\mathcal{O}_{\mathbb{C}^{n+k},x}.$$

We interrupt the discussion of this example to prove a very useful theorem, due to Rim and Teissier.

(4.8) *Theorem.* (Purity of the discriminant.) Let $f : (\mathbb{C}^{n+k},x) \to (\mathbb{C}^k,0)$, $k \geq 1$, define an icis of dim n. Then the ideal in $\mathcal{O}_{\mathbb{C}^k,0}$ defining the discriminant $(D_f,0)$ is principal.

Proof. Recall from (4.4) that $\mathcal{O}_{C,x}$ is a Cohen-Macaulay ring of dim k-1. Since $f : (C,x) \to (\mathbb{C}^k,0)$ is finite, $\mathcal{O}_{C,x}$ is a finite $\mathcal{O}_{\mathbb{C}^k,0}$-module and hence $\mathcal{O}_{C,x}$ is also Cohen-Macaulay as an $\mathcal{O}_{\mathbb{C}^k,0}$-module and $\operatorname{depth}_{\mathcal{O}_{\mathbb{C}^k,0}} \mathcal{O}_{C,x} =$ $= k-1$. (Serre (1975), Ch. IV, Prop.11, 12.) As $\mathcal{O}_{\mathbb{C}^k,0}$ is a k-dimensional regular local ring this implies that the homological dimension of $\mathcal{O}_{C,x}$ as an $\mathcal{O}_{\mathbb{C}^k,0}$-module is 1 (*op.cit.* Ch. IV, Prop. 21). In other words, there is an exact sequence of $\mathcal{O}_{\mathbb{C}^k,0}$-modules

$$0 \to \mathcal{O}^q_{\mathbb{C}^k,0} \xrightarrow{\alpha} \mathcal{O}^p_{\mathbb{C}^k,0} \to \mathcal{O}_{C,x} \to 0.$$

As α is injective, we must have $q \leq p$. On the other hand, $\mathcal{O}_{C,x}$ is supported by (C,x) and so $q \geq p$. Hence $q = p$, so that the 0th Fitting ideal of $\mathcal{O}_{C,x}$ (viewed as an $\mathcal{O}_{\mathbb{C}^k,0}$-module) is generated by the determinant of α.

Let us continue with example 5. By (4.8) the ideal defining $(D,0)$ is generated by a single element $\delta \in \mathcal{O}_{\mathbf{C}^k,0}$. The *multiplicity* of $(D,0)$ (or $\mathcal{O}_{D,0}$) is just the smallest integer ν with $\delta \in m^{\nu}_{\mathbf{C}^k,0}$. We denote it by $\nu(D,0)$ (or $\nu(\mathcal{O}_{D,0})$). The Taylor expansion of δ then begins with a nonzero homogeneous polynomial of degree $\nu(D,0)$. The cone defined by this polynomial is the tangent cone of $(D,0)$. Notice that the tangent cone is defined in the tangent space of $\mathbf{C}^k$ at 0, rather than in $\mathbf{C}^k$ itself. By a (linear) coordinate change we can always achieve that the last coordinate axis ℓ is not contained in this cone. Then $\delta(0,\ldots,0,t) = ct^{\nu(D,0)}$ + h.o.t. with $c \neq 0$. It follows that in this case that $\nu(D,0)$ is the $\mathbf{C}$-dim of

$$\mathcal{O}_{\mathbf{C}^{n+k},x}/(\{\frac{\partial(f_1,\ldots,f_k)}{\partial(z_{\nu_1},\ldots,z_{\nu_k})} : 1 \leq \nu_1 < \ldots < \nu_k \leq n+k\}, f_1,\ldots,f_{k-1})\mathcal{O}_{\mathbf{C}^{n+k},x}.$$

We close this chapter by investigating the geometric meaning of the condition that the discriminant space be a reduced hypersurface.

(4.9) *Lemma*. Let $f : X \to S$ be a good representative of a germ defining an icis. Then $s \in D$ is a regular point of D (by which we mean that $\mathcal{O}_{D,s}$ is a regular local ring - this is equivalent to: D is nonsingular in s and $\mathcal{O}_{D,s}$ is reduced) if and only if $C \cap X_s$ consists of a single quadratic singularity.

Proof. 'If' easily follows from (4.2)(iii) ⇒ (iv). To prove the 'only if' part, assume that s is a regular point of D. As an $\mathcal{O}_{S,s}$-module, $(f_*\mathcal{O}_C)_s$ is naturally isomorphic to $\oplus_{x\in C\cap X_s} f_*\mathcal{O}_{C,x}$, so the ideal defining D at s is the product of the Fitting ideals $F_0(f_*\mathcal{O}_{C,x})$. Since this ideal is prime, only one factor can occur. In other words, $C \cap X_s$ consists of a single point $\{x\}$. Following example 5 we have $1 = \nu(D,s)$ $\geq \dim_{\mathbf{C}}(\mathcal{O}_{X,x}/C_x + m_{S,s}\mathcal{O}_{X,x})$. Since $C_x + m_{S,s}\mathcal{O}_{X,x} \subset m_{X,x}$ this can only be when

$C_x + m_{S,s}O_{X,x} = m_{X,x}$. By (4.2)(ii) ⇒ (iii) this is equivalent to x being a quadratic singularity of f.

(4.10) *Corollary*. Let $f : (\mathbb{C}^{n+k},x) \to (\mathbb{C}^k,0)$, define an icis of dim n. Then $O_{D,0}$ is reduced if and only if f admits a good representative $f : X \to S$ such that the set of $s \in D$ for which $C \cap X_s$ consists of a single quadratic singularity is dense in D.

Proof. Choose a good representative $f : X \to S$ such that its discriminant space D is defined by a single element $\delta \in \Gamma(S,O_S)$. By shrinking S we can assume that a prime factorization $\delta = \delta_1^{\nu_1}\ldots\delta_\ell^{\nu_\ell}$ of δ in $O_{\mathbb{C}^k,0}$ corresponds to a decomposition $D_1,\ldots,D_\ell$. Now $O_{D,0}$ is reduced if and only if all the exponents $\nu_1,\ldots,\nu_\ell$ are 1. The corollary then follows from (4.9).

(4.11) *Corollary*. Let $f : (\mathbb{C}^{n+k},x) \to (\mathbb{C}^k,0)$ define an icis of dim n and suppose that f is Thom-transversal. Then f has reduced and irreducible discriminant space.

Proof. This follows from (4.7) and (4.10).

With the help of the last corollary we show that we can always 'embed' a map-germ defining an icis in a map-germ having reduced discriminant, a result which will prove its usefulness in the next chapter.

(4.12) *Proposition*. Let $f : (\mathbb{C}^{n+k},x) \to (\mathbb{C}^k,0)$ define an icis of dim n. Then there exists an analytic germ $F : (\mathbb{C}^{n+k}\times\mathbb{C}^\ell,x\times 0) \to (\mathbb{C}^k\times\mathbb{C}^\ell,0)$ of the form $F(z,u) = (g(z,u),u)$ ($z \in \mathbb{C}^{n+k}$, $u \in \mathbb{C}^\ell$) with $g(z,0) = f(z)$ such that F is Thom-transversal. In particular F has reduced and irreducible discriminant.

Proof. Take $\ell = (n+k)k$ and identify $\mathbb{C}^\ell$ with $\mathrm{Hom}(\mathbb{C}^{n+k},\mathbb{C}^k)$. Then

$F(z,u) = (f(z)+u(z),u)$ is as required.

Notice that in this proposition f and F make up a cartesian diagram. So if δ is a prime element of $\mathcal{O}_{\mathbb{C}^{k+\ell},x\times 0}$ which defines the discriminant space of F, then $\delta|(\mathbb{C}^k,x)$ defines the discriminant space of f.

5 RELATIVE MONODROMY

The term relative monodromy refers to a powerful method by means of which many homotopy properties of Milnor fibres and Milnor fibrations can be proved inductively. As far as the author knows this notion was introduced by R. Thom but the most effective use of it was certainly made by D.T. Lê (1973, 1978).

The usual situation is that one is given an analytic germ $f = (f_1,\ldots,f_k) : (X,x) \to (\mathbb{C}^k,0)$ such that both f and $f' := (f_1,\ldots,f_{k-1}) : (X,x) \to (\mathbb{C}^{k-1},0)$ define an isolated singularity. A basic result then asserts that f admits a good representative of the form $f : X \to S'\times\Delta$ $(\subset \mathbb{C}^{k-1}\times\mathbb{C})$ such that its composition with the projection $S'\times\Delta \to S'$ is topologically equivalent to a good representative of f'. This fact is used to prove many homotopy properties of Milnor fibrations inductively: (a) with induction on the fibre dimension n that any fibre has the homotopy type of a finite complex of dim $\leq$ n, (b) essentially with induction on k that in the complete intersection case any fibre of dim n is (n-1)-connected and finally (c), if k = 1 (with induction on n again) that the homological monodromy of the Milnor fibration is quasi-unipotent of index $\leq$ n.

5.A *The basic construction*

(5.1) Throughout this section X stands for an analytic set in an open $U \subset \mathbb{C}^N$ and x is a point of X such that $X-\{x\}$ is nonsingular of pure dimension $n+k$. Suppose we have an analytic map-germ $f : (X,x) \to (\mathbb{C}^k,0)$ defining an isolated singularity. Then according to (2.8) the discriminant locus $(D,0)$ of f will be a hypersurface germ. By a coordinate change in $\mathbb{C}^k$ we can arrange that the last coordinate axis ℓ of $\mathbb{C}^k$ meets $(D,0)$ at most in $\{0\}$. This property is equivalent to the condition that $f' := (f_1,\ldots,f_{k-1}) : (X,x) \to (\mathbb{C}^{k-1},0)$ defines an isolated singularity. We show that we can also go in the 'opposite direction', see (5.3) below.

(5.2) *Lemma*. Let Y be an analytic set at $0 \in \mathbb{C}^N$ such that $Y-\{0\}$ is nonsingular of pure dimension d. Then for a generic linear hyperplane H of $\mathbb{C}^N$, H intersects X transversally in a punctured neighbourhood of $0 \in \mathbb{C}^N$.

Proof. We use the *development* $\pi : \tilde{Y} \to Y$ of Y. Let Y^* denote the set of $(y,\tau) \in Y \times G_d(\mathbb{C}^N)$ with τ contained in the Zariski tangent space of Y. Using local coordinates we see that Y^* is an analytic subvariety of $Y \times G_d(\mathbb{C}^N)$: if $\phi_1,\ldots,\phi_\ell$ generate I_Y in $V \subset \mathbb{C}^N$ then $Y^* \cap (V \times G_d(\mathbb{C}^N))$ consists of the $(y,\tau) \in (Y \cap V) \times G_d(\mathbb{C}^N)$ with $d\phi_\lambda|\tau = 0$ $(\lambda=1,\ldots,\ell)$. We let $\tilde{Y}$ be the union of the irreducible components of Y^* which contain a point of the form (y,T_yY) with Y smooth at y of dim d, and let $\pi : \tilde{Y} \to Y$ be the projection. As $\pi^{-1}(0)$ is a proper subvariety of $\tilde{Y}$, it will have dim $\leq d-1$. Now the hyperplanes of $\mathbb{C}^N$ containing a fixed d-dimensional subspace form a projective space of dim $N-d-1$. So the hyperplanes containing a subspace corresponding to a point of $\pi^{-1}(0)$ form a subvariety of $\check{\mathbb{P}}^{N-1}$ of dim $\leq (N-d-1) + (d-1) = N-2$. Any $H \in \check{\mathbb{P}}^{N-1}$ outside this subvariety has the desired property.

(5.3) *Corollary*. In the situation of (5.1) there exists an $f_{k+1} \in m_{X,x}$

such that $g := (f, f_{k+1}) : (X,x) \to (\mathbf{C}^{k+1}, 0)$ defines an isolated singularity and if $k > 0$ g admits a representative such that the quadratic singularities are dense in its critical locus.

Proof. According to the previous lemma there exists a $\phi_0 \in \mathrm{Hom}(\mathbf{C}^N, \mathbf{C})$ such that the affine hyperplane defined by $\phi_0 - \phi_0(x)$ meets (X,x) transversally in a punctured neighbourhood of x. Then there is a good proper representative $f : \overline{X} \to S$ of f such that the critical locus of $(f, \phi_0) : \overline{X} \to S \times \mathbf{C}$ doesn't meet $\partial\overline{X}$. Let U be an open neighbourhood of ϕ_0 in $\mathrm{Hom}(\mathbf{C}^N, \mathbf{C})$ such that all $\phi \in U$ possess this property. If $k > 0$ we fix a $t \in S - D_f$ (so that X_t is regular) and consider the map $F : \overline{X}_t \times U \to \mathbf{C} \times U$, $F(z, \phi) = (\phi(z), \phi)$. It is easy to check that the critical space C_F of F is regular. By the very choice of U, C_F doesn't meet $\partial\overline{X}_S \times U$ and is therefore finite over U. Let $\phi \in U$ be a regular value of $\pi_U | C$. Then it follows from (4.2) that $\phi | \overline{X}_t$ has only quadratic singularities. Every irreducible component C_1 of the critical locus of $(f, \phi) : \overline{X} \to S \times \mathbf{C}$ is finite over S and of the same dimension as S. Hence C_1 maps onto S, in particular, $C_1 \cap X_t \neq \emptyset$. Since C is reduced in $C \cap \overline{X}_t$, it follows that C is generically reduced. This means that the quadratic singularities are dense in C. So if we take for f_{k+1} the image of $\phi - \phi(x)$ in $\mathcal{O}_{X,x}$, then f_{k+1} is as required.

From now on we suppose we are given an analytic map $f : X \to \mathbf{C}^k$ with $f(x) = 0$ such that both f and $f' = (f_1, \ldots, f_{k-1})$ define an isolated singularity at x. The crucial result of this section is

(5.4) *Proposition*. There exist pairs of good proper representatives of f and f' of the form $f : \overline{X} \to S = S' \times \overline{\Delta}$, $f' : \overline{X'} \to S'$ ($S' \subset \mathbf{C}^{k-1}$ open, Δ a disc in $\mathbf{C}$ centered at 0) with $\overline{X} \subset \overline{X'}$, a closed disc $\overline{\Delta}_1 \subset \Delta$ such that

$S'\times\Delta_1$ contains the discriminant of f and a homeomorphism H of $\overline{X}$ onto $\overline{X}'$ such that $f'\circ H = \pi\circ f$ ($\pi : S'\times\overline{\Delta} \to S'$ being the projection) and H is the identity on $f^{-1}(S'\times\overline{\Delta}_1)$.

Proof. We begin with choosing $\varepsilon > 0$, $\eta > \eta_1 > 0$ and $\eta' > 0$ such that $f : \overline{X} \to S'\times\overline{\Delta}$ and $f' : \overline{X}' \to S'$ are good representatives and $D_f \subset S'\times\overline{\Delta}_1$. Here

$$\overline{X} = \{z \in X : |z-x| \le \varepsilon, \ |f_k(z)| \le \eta, \ |f'(z)| < \eta'\}$$
$$\overline{X}' = \{z \in X : |z-x| \le \varepsilon, \ |f'(z)| < \eta'\}$$
$$S' = \{t \in \mathbf{C}^{k-1} : |t| < \eta'\}$$

and Δ resp. Δ_1 is the open disc in $\mathbf{C}$ centered at 0 of radius η resp. η_1. Notice that $\overline{X} \subset \overline{X}'$. In the course of the proof we decrease these numbers whenever necessary.

Claim. If $\varepsilon > 0$ is sufficiently small, the differentials of the functions $z \in \overline{X}'_0 - \overline{X}_0 \to |z-x|^2$, $z \in \overline{X}'_0 - \overline{X}_0 \to |f_k(z)|^2$ are nonzero and nowhere point in opposite directions.

If not, then by the curve selection lemma (2.1), there exists a real analytic curve $\gamma : [0,\varepsilon) \to X'_0$ with $\gamma(0) = x$ such that for $\tau \neq 0$ $\gamma(\tau) \notin X_0$ and the two differentials are at $\gamma(\tau)$ proportional by a non-positive factor. Since $|\gamma|^2$ is strictly increasing at x, $|f_k\circ\gamma|^2$ must then be nondecreasing at 0, which can only be if γ stays in X_0. This gives a contradiction.

Choose $\varepsilon > 0$ as in the claim. Put $Y := \overline{X}'-\overline{X}_{S'\times\Delta} = \{z \in \overline{X}' : |f_k(z)| \ge \eta\}$. Then Y is proper over S'. Since f' is a submersion on the compact set Y_0 ($\subset \overline{X}'_0-\{x\}$), we can choose $\eta' > 0$ so small that the critical locus of f avoids Y and the differentials of $z \in Y_{t'} \to |z-x|^2$,

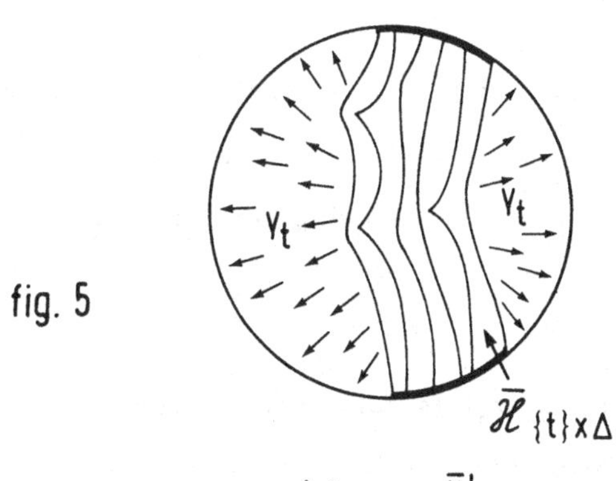

$z \in Y_{t'} \to |f_k(z)|^2$ never point in opposite directions for all $t \in S'$. This enables us to construct a C^∞ vector field v on $\mathcal{Y}$ such that $df_1,\ldots,df_{k-1}$ annihilate v, and the differentials of $z \to |z-x|^2$ and $z \to |f_k(z)|^2$ are > 0 on v. So $|f_k|$ and $|z-x|$ strictly increase along integral curves γ of v whereas $f_1,\ldots,f_{k-1}$ are constant on them. In particular, any integral curve of v extends to one which starts in $\bar{X}_{S'\times\partial\Delta} = \{z \in \bar{X}' : |f_k(z)| = \eta\}$, leaves $\bar{X}$ in a point of $\partial\bar{X}'$ and remains inside a fibre of f'. If $z \in \bar{X}_{S'\times\partial\Delta}$, let $\tau(z) \in [0,\infty)$ be the first time that the integral curve γ_z through z hits $\partial\bar{X}'$. Then τ is a continuous function and $\tau(z) = 0$ if and only if $z \in \partial\bar{X}_{S\times\partial\Delta}$. The map $(z,t) \to \gamma_z(t)$ identifies $\mathcal{Y}$ with $\{(z,t) \in \bar{X}_{S'\times\partial\Delta} : 0 \le t \le \tau(z)\}$. This identification can be used to construct in a rather straightforward manner a homeomorphism H of $\bar{X}$ onto $\bar{X}'$ which commutes with f' and is the identity over $S'\times\bar{\Delta}_1$.

5.B *The homotopy type of the Milnor fibre*

We continue with the situation described in (5.4). Our purpose is to compare the two Milnor fibrations f and f'. Let $t \in S'$ be arbitrary and set $s = (t,\eta) \in S' \times \overline{\Delta}$ (recall that η is the radius of Δ). By prop. (5.4), $\pi \circ f : \overline{X} \to S'$ is topologically equivalent to a good representative of f'. Our first goal is to prove the following two assertions simultaneously.

$(5.5)_{n+1}$ *Assertion*. The pair $(\overline{X}'_t, \overline{X}_s)$ is relatively homotopic to a relative finite cell complex of dim $\leq n+1$.

$(5.6)_n$ *Assertion*. Any fibre of a good representative of a map-germ defining an isolated singularity of complex dim n is homotopy equivalent to a finite cell complex of dim $\leq n$.

The last assertion is actually a special case of a result due to Andreotti & Narasimhan which says that any Stein space of dim n with only isolated singularities has the homotopy type of a CW complex of dim $\leq n$.

Proofs. It is clear that $(5.5)_{n+1}$ and $(5.6)_n$ imply $(5.6)_{n+1}$. We show that $(5.6)_n$ implies $(5.5)_{n+1}$. Since $(5.6)_0$ is trivially true both assertions then follow with induction on n.

So let $s_1, \dots, s_m$ denote the distinct points of $D_f \cap (\{t\} \times \overline{\Delta})$ and choose about each s_μ a closed disc $\overline{\Delta}_\mu$ in $\{t\} \times \Delta$ such that for each singular point $x_{\mu,\alpha} \in X_{s_\mu}$, the germ $f|(X'_t, x_{\mu,\alpha})$ admits a good proper representative over $\overline{\Delta}_\mu$: $f : \overline{X}_{\mu,\alpha} \to \overline{\Delta}_\mu$. We let the $\overline{\Delta}_\mu$'s and the $\overline{X}_{\mu,\alpha}$'s be disjoint. Let Γ be a tree in $\{t\} \times \overline{\Delta}$ such that Γ meets $\overline{\Delta}_\mu$ in a closed 1-simplex $[s_\mu, s'_\mu]$ $(\mu = 1, \dots, m)$ and Γ contains s. Put $A := \cup_\mu \overline{\Delta}_\mu \cup \Gamma$ and $B := \Gamma - \cup_\mu (\Delta_\mu \cap \Gamma)$. Then A resp. $\{s\}$ is a strong deformation retract of $\{t\} \times \overline{\Delta}$ resp. B. Since f is locally trivial over $\{t\} \times \Delta - \cup_\mu \{s_\mu\}$, it follows from the homotopy lifting property that $\overline{X}_A$ resp. $\overline{X}_s$ is a strong deformation retract of $\overline{X}_{\{t\}\times\overline{\Delta}} = \overline{X}'_t$ resp. $\overline{X}_B$. For the same reason $\cup_{\mu,\alpha}\overline{X}_{\mu,\alpha} \cup \overline{X}_B$ is a strong deformation retract of $\overline{X}_A$. Put $F_{\mu,\alpha} := \overline{X}_B \cap \overline{X}_{\mu,\alpha}$. The space $\overline{X}_{\mu,\alpha}$ is a cone by (2.10). So the space $E' := \overline{X}_B \cup \cup_{\mu,\alpha}\mathrm{Cone}(F_{\mu,\alpha})$ obtained from $\overline{X}_B$ by putting a cone over each $F_{\mu,\alpha}$ sits in $E := \overline{X}_B \cup \cup_{\mu,\alpha}\overline{X}_{\mu,\alpha}$. In Whitehead's terminology, $(\cup_{\mu,\alpha}\overline{X}_{\mu,\alpha}, \cup_{\mu,\alpha}F_{\mu,\alpha})$ and (E',E) are NDR-pairs and so E' is a strong deformation retract of E (Whitehead (1978), (I.5.12) and (I.5.9)). If $r : \overline{X}_B \to \overline{X}_s$ trivializes $f : \overline{X}_B \to B$ (with $r|\overline{X}_s = 1$), then r induces a map from E' to the space $E'' := \overline{X}_s \cup \cup_{\mu,\alpha}\mathrm{Cone}(F_{\mu,\alpha})$. obtained from $\overline{X}_s$ by putting a cone over each $r(F_{\mu,\alpha})$ and this map is easily seen to be a homotopy equivalence rel. $\overline{X}_s$. Combining these homotopy equivalences we find that $(\overline{X}'_t,\overline{X}_s)$ is relatively homotopy equivalent to $(E'',\overline{X}_s)$. Now $(5.5)_{n+1}$ follows, since by $(5.6)_n$ each $F_{\mu,\alpha}$ has the homotopy type of a finite cell complex of dim $\leq n$.

(5.7) An important case is when all the singular points are quadratic. Then according to (3.A) each $F_{\mu,\alpha}$ has an n-sphere $S^n_{\mu,\alpha}$ as a deformation retract. It follows from the above proof that up to relative homotopy, $\overline{X}'_t$ is obtained from $\overline{X}_s$ by attaching to $\overline{X}_s$ (n+1)-cells: one for each singular point $x_{\mu,\alpha}$, with attaching map $h|S^n_{\mu,\alpha}$.

We now concentrate on the complete intersection case. We are going to prove

(5.8) The Milnor fibre of an icis of dim n is (n-1)-connected.

In the hypersurface case this result is due to Milnor (1968). It was extended to complete intersections by Hamm (1971a,b) using Morse functions. (Milnor fibres of isolated singularities which are not complete intersections may have nontrivial homotopy groups in lower dimensions, see Pinkham (1974), Wahl (1981) and Greuel & Steenbrink (1983)).

The proof will use induction on the embedding codimension of (X_0,x) (see (1.9)). We want the following

(5.9) *Lemma*. Let $f : (\mathbb{C}^{n+k},x) \to (\mathbb{C}^k,0)$ define an icis of dim n. If a line ℓ in $\mathbb{C}^k$ (passing through 0) is not in the image of $df(x)$, then the embedding codimension of $(f^{-1}(\ell),x)$ is one less than the embedding codimension of $(f^{-1}(0),x)$.

Proof. Let r denote the rank of $df(x)$. Since ℓ is not in the image of $df(x)$, we may assume that after a linear coordinate change in $\mathbb{C}^k$, $df_1(x),\dots,df_r(x)$ are linearly independent, $df_{r+1}(x) = \dots = df_k(x) = 0$ and ℓ is the last coordinate axis. Then $(f^{-1}(\ell),x)$ is defined by $(f_1,\dots,f_{k-1}) : (\mathbb{C}^{n+k},x) \to (\mathbb{C}^{k-1},0)$ so that its embedding codimension equals $(k-1)-r$.

Now let $f : (\mathbb{C}^{n+k},x) \to (\mathbb{C}^k,0)$ define an icis of dim n. In view of (2.9) it suffices to prove (5.8) for just one good representative of f. This enables us to assume that f has reduced discriminant. For f can always be extended to an analytic germ $F : (\mathbb{C}^{n+k+\ell},(x,0)) \to (\mathbb{C}^{k+\ell},0)$ of the form $F(z,u) = (g(z,u),u)$, $g(z,0) = f(z)$ with reduced discriminant by (4.12) and a good representative of F will restrict to a good representative of f (so that both have isomorphic Milnor fibres). We now proceed with the embedding codimension of $(f^{-1}(0),x)$. If it is zero, then f is a submersion and so there is nothing to prove. If not, then lemma (5.9) allows us to assume (by a possible coordinate change in $\mathbb{C}^k$) that $f' = (f_1,\dots,f_{k-1}) : (\mathbb{C}^{n+k},x) \to (\mathbb{C}^{k-1},0)$ defines an icis of dim n+1 of lower embedding codimension than $(f^{-1}(0),x)$. Now choose a good proper representative $f : \overline{X} \to S'\times\overline{\Delta}$ as in prop. (5.4). Let $t \in S'$ be such that t is not in the discriminant of $\pi\circ f$ and $\{t\}\times\overline{\Delta}$ meets the discriminant of f in regular points $s_1,\ldots,s_m$ only. Over each s_μ we then have exactly one singular point and this point is quadratic. Now $\overline{X}'_t$ is homeomorphic to a Milnor fibre of f' and so by our inductive assumption n-connected. In (5.7) we found that up to homotopy, $\overline{X}'_t$ is obtained from $\overline{X}_s$ by attaching m (n+1)-cells to it, so that the pair $(\overline{X}'_t,\overline{X}_s)$ is n-connected (e.g. Whitehead (1978), (II.3.9)). It follows that $\overline{X}_s$ is (n-1)-connected.

(5.10) *Corollary*. Let $f : \overline{X} \to S$ be a good proper representative of a germ $f : (\mathbb{C}^{n+k},x) \to (\mathbb{C}^k,0)$ defining an icis of dim n. Then every fibre $\overline{X}_s$ has the homotopy type of a finite *bouquet of n-spheres* $S^n \vee\ldots\vee S^n$ (i.e. a finite union of n-spheres having one point in common).

Proof. We first show that an (n-1)-connected finite cell complex Y of dim $\leq$ n is homotopy equivalent to a finite bouquet of spheres. For $n = 1$ the assertion is obvious. If $n \geq 2$, the Hurewicz map $\pi_n(Y,\text{base point}) \to H_n(Y)$ will be an isomorphism (by the Hurewicz theorem). The torsion subgroup of $H_n(Y)$ is isomorphic to the torsion subgroup of $H^{n+1}(Y)$ and hence trivial. So we can choose finitely many maps $(S^n,\text{base point}) \to (Y,\text{base point})$, whose homotopy classes yield a basis for $\pi_n(Y,\text{base point})$. These combine to give a map from $S^n \vee\ldots\vee S^n$ to Y inducing an isomorphism on homology. Whitehead's theorem then implies that this is a homotopy equivalence.

If $s \in S-D$, then the corollary is now immediate from (5.5) and (5.7). If $s \in D$, then, as the argument used in the proof of (5.5) shows, there exists an $s_1 \in S-D$ close to s such that $\overline{X}_s$ is homotopy equivalent to the space obtained from $\overline{X}_{s_1}$ by attaching a cone over each small Milnor fibre in X_{s_1} corresponding to a singular point of $\overline{X}_s$. As such a Milnor fibre is homotopy equivalent to a bouquet of n-spheres, this amounts to attaching (n+1)-cells to $\overline{X}_{s_1}$. This implies that X_s is also (n-1)-connected. The corollary then follows from (5.5) and the above assertion.

(5.11) The number of n-spheres occurring in the bouquet describing the homotopy type of a Milnor fibre of an n-dimensional icis (X_0,x) is called the *Milnor number* and denoted $\mu(X_0,x)$. This in anticipation of something that will be proved in the next chapter, namely that any two Milnor fibres of (X_0,x) are diffeomorphic. Notice that $\mu(X_0,x) = \dim \widetilde{H}_n(\overline{X}_s,\mathbb{R})$. In order to compute this number, we return to the proof of (5.7). The exact (reduced) homology sequence of the pair $(\overline{X}'_t,\overline{X}_s)$ is by (5.10)

$$.. \to 0 \to H_{n+1}(\overline{X}'_t) \to H_n(\overline{X}'_t,\overline{X}_s) \overset{\partial_*}{\to} \widetilde{H}_n(\overline{X}_s) \to 0 \to ..$$

where the possibly nonzero terms are free $\mathbb{Z}$-modules of rank $\mu(X'_0,x)$, m and $\mu(X_0,x)$ respectively. So $m = \mu(X_0,x) + \mu(X'_0,x)$. If we orient each vanishing sphere $S_\mu \subset \overline{X}_s$ ($\mu=1,\ldots,m$) then the attached (n+1)-cells along these spheres correspond to a basis for $H_{n+1}(\overline{X}'_t,\overline{X}_s)$. The image of this basis under ∂_* is the set of vanishing cycles $\delta_\mu := [S_\mu]$ ($\mu=1,\ldots,m$), which therefore *generates* $\widetilde{H}_n(\overline{X}_s)$. The case when (X_0,x) is a hypersurface is especially noteworthy: then its embedding codimension is 1, so that (X'_0,x) must have embedding codimension 0 (i.e. f' is a submersion). It follows that ∂_* is an isomorphism and that $\delta_1,\ldots,\delta_m$ is a *basis* of $\widetilde{H}_n(\overline{X}_s)$. Returning to the general case, we recall that m was defined as the number of intersection points of D_f with the line $\{t\}\times\mathbb{C}^{k-1}$ parallel to ℓ. Each intersection point has multiplicity one, so m is given by the formula in example 5 of Ch. 4. Hence

$$(5.11.a)\qquad \mu(X_0,x) + \mu(X_0',x) = \dim_{\mathbb{C}} \mathcal{O}_{\mathbb{C}^{n+k},x} / (\{\frac{\partial(f_1,\ldots,f_k)}{\partial(z_{\nu_1},\ldots,z_{\nu_k})} : 1 \le \nu_1 <\ldots< \nu_k \le n+k\}, f_1,\ldots,f_{k-1})\mathcal{O}_{\mathbb{C}^{n+k},x}.$$

This formula is proved here under the assumption that D_f is reduced and $(f_1,\ldots,f_{k-1})$ defines an icis of dim n+1. But we claim its validity whenever the right hand side is finite: the finiteness assumption just tells us that the critical space of $f_k : (X_0',x) \to (\mathbb{C},0)$ is concentrated in $\{x\}$ and as far as the reducedness of the discriminant is concerned: as in the proof of (5.8) we may extend f to a germ F with reduced discriminant and observe that the corresponding terms in the formula (5.11.a) are the same for F and f. Notice that in the hypersurface case (5.11.a) simply reduces to

$$(5.11.b)\qquad \mu(X_0,x) = \dim_{\mathbb{C}} \mathcal{O}_{\mathbb{C}^{n+1},x} / (\frac{\partial f}{\partial z_1},\ldots,\frac{\partial f}{\partial z_{n+1}})\mathcal{O}_{\mathbb{C}^{n+1},x}.$$

In the general case we get (by taking alternating sums of formulas of type (a)):

$$(5.11.c)\qquad \mu(X_0,x) = \Sigma_{\kappa=1}^{k}(-1)^{k-\kappa}\dim_{\mathbb{C}} A_\kappa,\ \text{with}$$

$$A_\kappa = \mathcal{O}_{\mathbb{C}^{n+k},x} / (\{\frac{\partial(f_1,\ldots,f_\kappa)}{\partial(z_{\nu_1},\ldots,z_{\nu_\kappa})}: 1 \le \nu_1 <\ldots< \nu_\kappa \le n+k\}, f_1,\ldots,f_{\kappa-1})\mathcal{O}_{\mathbb{C}^{n+k},x},$$

valid, whenever all the A_κ's are finite dimensional.

Example 1. Let $\lambda_1,\ldots,\lambda_{n+2}$ be distinct complex numbers. Then $f : (\mathbb{C}^{n+2},0) \to (\mathbb{C}^2,0)$, $z \to (z_1^2+\ldots+z_{n+2}^2, \lambda_1 z_1^2+\ldots+\lambda_{n+2}z_{n+2}^2)$ defines an icis (X_0,x) of dim n. Consider the ideal

$$I := (\{\frac{\partial(f_1,f_2)}{\partial(z_i,z_j)} : 1 \le i < j \le n+2\}, f_1)\mathcal{O}_{\mathbb{C}^{n+2},0}$$

$$= (\{z_i z_j : 1 \le i < j \le n+2\}, z_1^2+\ldots+z_{n+2}^2)\mathcal{O}_{\mathbb{C}^{n+2},0}.$$

This ideal contains all homogeneous polynomials of degree 3, so that $I + m^4_{\mathbb{C}^{n+2},0} \supset m^3_{\mathbb{C}^{n+2},0}$. It then follows from Nakayama's lemma that $I \supset m^3_{\mathbb{C}^{n+2},0}$. This implies that $\dim_{\mathbb{C}}\mathcal{O}_{\mathbb{C}^{n+2},0}/I$ is finite and makes the latter easy to compute: the residue classes of $1, z_1, \ldots, z_{n+2}, z_1^2, \ldots, z_{n+1}^2$ form a basis of $\mathcal{O}_{\mathbb{C}^{n+2},0}/I$, so that its dimension is $2(n+2)$. The dimension of

$\mathcal{O}_{\mathbb{C}^{n+2},0}/(\frac{\partial f_1}{\partial z_1},\ldots,\frac{\partial f_1}{\partial z_{n+2}})\mathcal{O}_{\mathbb{C}^{n+2},0} = \mathcal{O}_{\mathbb{C}^{n+2},0}/m_{\mathbb{C}^{n+2},0}$ equals 1 (this also follows from the fact that this computes the Milnor number of a quadratic singularity) and so $\mu(X_0,0) = 2(n+2)-1 = 2n+3$.

Formula (5.11.c) is difficult to work with if the number k is large. In the extreme case of a 0-dimensional icis, there is however a simple alternative formula.

(5.12) *Proposition*. Let $f : (\mathbb{C}^k,0) \to (\mathbb{C}^k,0)$ define an icis. Then $\mu(X_0,0) = \deg(f)-1$, where $\deg(f)$ is the degree of f, given by the $\mathbb{C}$-dimension of $\mathcal{O}_{\mathbb{C}^k,0}/(f_1,\ldots,f_k)\mathcal{O}_{\mathbb{C}^k,0}$.

Proof. Let $f : X \to S$ be a good representative. Then $f_*\mathcal{O}_X$ is a coherent $\mathcal{O}_S$-module whose stalk in $0 \in S$ is $\mathcal{O}_{X,0}$. As an $\mathcal{O}_{S,0}$-module $\mathcal{O}_{X,0}$ is free (Serre (1975), Ch. IV, Cor. to Prop. 22), so by shrinking S we may as well assume that $f_*\mathcal{O}_X$ is a free $\mathcal{O}_S$-module. The 'fibre dimension' of $f_*\mathcal{O}_X$ in $s \in S-D$ is just the cardinal of $f^{-1}(s)$ (which is $\mu(X_0,0)+1$). If we evaluate this dimension in $0 \in S$, we find that it also equals $\dim_{\mathbb{C}}\mathcal{O}_{\mathbb{C}^k,0}/(f_1,\ldots,f_k)\mathcal{O}_{\mathbb{C}^k,0}$.

5.C *The monodromy theorem*

If N is a nilpotent endomorphism of a complex vector space V then $\log(1+N) = \Sigma_{k=1}^{\infty} -\frac{1}{k}(-N)^k$ is a polynomial in N and we can write $\log(1+N) = N.U = U.N$ with $U := \Sigma_{k=0}^{\infty} \frac{1}{k+1}(-N)^k$. Since U is of the form $1 + M$, with M nilpotent, U is invertible: $U^{-1} = \Sigma_{k=0}^{\infty}(-M)^k$. We say that N is *nilpotent of index* $\le m$ if $N^{m+1} = 0$. It is clear that this is so if and only if $(\log(1+N))^{m+1} = 0$. Similarly, we say that an endomorphism X of V is *unipotent*, resp. *quasi-unipotent of index* $\le m$ if $X-1$, resp. X^k-1 for some $k \in N$, is nilpotent of index $\le m$. This implies that X is invertible. In fact, in case $\dim V < \infty$, the second condition just means that the eigenvalues of X are roots of unity and that the Jordan blocks of X have size $\le m+1$. A set of automorphisms of V is called (quasi-) unipotent of index $\le m$ if each member is. Usually we abbreviate quasi-unipotent by q.u.

The following simple lemma will prove quite useful.

(5.13) *Lemma*. For a pair of commuting unipotent automorphisms X,Y of a complex vector space V the following are equivalent:

(i) The semigroup generated by X and Y in Aut(V) is unipotent of index $\le m$.

(ii) The group generated by X and Y in Aut(V) is unipotent of index $\le m$.

(iii) $(X-1)^p(Y-1)^{m+1-p} = 0$ for $p = 0,\dots,m+1$.

Proof. Write $N = \log X$, $M = \log Y$. Then $NM = MN$ and $\log(X^pY^q) = pN+qM$. So the lemma asserts the equivalence of (i) $(pN+qM)^{m+1} = 0$ for $(p,q) \in \mathbb{Z}_+^2$, (ii) $(pN+qM)^{m+1} = 0$ for $(p,q) \in \mathbb{Z}^2$ and (iii) $(pN+qM)^{m+1} = 0$ for $p+q = m+1$, $p = 0,\dots,m+1$. This is a simple exercise in $\mathbb{Q}[N,M]$ which we leave to the reader.

Quasi-unipotent automorphisms often arise in algebraic geometry as monodromy transformations. For instance, a theorem due to Clemens (1969) and Landman (1973) asserts that if $f : X \to \Delta$ is a proper analytic map from an analytic manifold of dim n+1 to a complex disc at $0 \in \mathbb{C}$ which is of maximal rank over $\Delta^* := \Delta - \{0\}$ and $s \in \Delta^*$, then $\pi(\Delta^*,s)$ acts on $H^p(X_s,\mathbb{C})$ as a group of q.u. automorphisms of index $\leq \min\{p,2n-p\}$. We will be concerned with the following local analogue of this.

(5.14) *Theorem*. Let (X,x) be an isolated singularity of dim n+1 and let $f \in m_{X,x}$, not a zero-divisor. Then there is a good proper representative $f : \overline{X} \to \Delta$ of $f : (X,x) \to (\mathbb{C},0)$ such that for $s \in \Delta^*$ the action of $\pi(\Delta^*,s)$ on $H^p(\overline{X}_s,\mathbb{C})$ is trivial if $p \neq n$ and q.u. of index $\leq n$ for $p = n$.

The Clemens-Landman proof can be adapted to give a proof of (5.14) (D. Fried, to appear). A crucial ingredient of their proof is Hironaka's resolution theorem. Lê (1978), using relative monodromy and his so-called 'carrousel method', gave a more geometric proof which avoided the resolution of singularities, but did not yield the index estimate. In a Bourbaki talk Lê (1981) announced that his method combined with ideas of Nilsson gives this estimate as well. The proof given here is inspired by Lê's carrousel method.

Our aim is to reduce the proposition to an assertion about sheaves on a two-dimensional polycylinder. For this we recall that if Z is a space, z_0 a point of Z and F a local system on Z (i.e. a locally constant sheaf of abelian groups on Z), then we have a natural (monodromy) action of $\pi(Z,z_0)$ on the stalk F_{z_0}. If F is a local system of complex vector spaces over Z, then we say that F has *q.u. monodromy of index* $\leq$ m if $\pi(Z,z)$ acts as such on F_z for all $z \in Z$. If Z is arcwise connected, then the action of $\pi(Z,z_0)$ on F_{z_0} completely determines F up to canonical isomorphism and we only need to verify the quasi-unipotence in z_0. For instance, if $R^q f_*(\mathbb{C}_{\overline{X}})$ denotes the sheaf on Δ associated to $\Delta \supset V \to H^q(f^{-1}(V);\mathbb{C})$, then the theorem says that $R^q f_*(\mathbb{C}_{\overline{X}})|\Delta^*$ has trivial monodromy for $q \neq n$ and q.u. monodromy of index $\leq n$ if $q = n$.

Proof of (5.14). We proceed with induction on n. For $n = 0$, the assertion is trivial and so we assume $n > 0$. Corollary (5.3) and the construction of (5.4) furnish an analytic map $f : \overline{X} \to \Delta$ which is topologically equivalent to a good proper representative of f and a factorization

$$f : \overline{X} \xrightarrow{g} \Delta\times\overline{\Delta}' \xrightarrow{\pi} \Delta$$

(where Δ and Δ' are complex discs) such that $g^{-1}(0,0)-\{x\}$ is regular, g is a good representative of its germ at $(0,0)$, the discriminant D of g is contained in $\Delta\times\Delta'$ and the quadratic singularities are dense in the critical locus of g. By shrinking Δ, we may assume that either $C \subset \{x\}$ or that $C-\{x\}$ consists of quadratic singularities and $g : C-\{x\} \to D-\{(0,0)\}$, $\pi : D-\{(0,0)\} \to \Delta^*$ are unramified coverings.

Now let $(s,t) \in \Delta^*\times\partial\overline{\Delta}'$ and consider the cohomology sequence of the pair $(f^{-1}(s),g^{-1}(s,t))$:

$$..\to H^k(f^{-1}(s),g^{-1}(s,t);\mathbb{C}) \to H^k(f^{-1}(s);\mathbb{C}) \to H^k(g^{-1}(s,t);\mathbb{C}) \to..$$

Notice that $\pi(\Delta^*,s)$ acts on this sequence. The action of $\pi(\Delta^*,s)$ on $H^k(g^{-1}(s,t);\mathbb{C})$ is trivial, simply because g is C^∞-trivial over $\Delta\times\{t\}$. Since $H^k(f^{-1}(s),g^{-1}(s,t);\mathbb{C}) = 0$ for $k \neq n$ (by (5.7)) it follows that $\pi(\Delta^*,s)$ acts trivially on $H^k(f^{-1}(s);\mathbb{C})$ for $k < n$. For $k > n$, $H^k(f^{-1}(s);\mathbb{C}) = 0$ and

$$H^n(f^{-1}(s),g^{-1}(s,t);\mathbb{C}) \to H^n(f^{-1}(s);\mathbb{C})$$

is surjective. So it suffices to show that the action of $\pi(\Delta^*,s)$ on $H^n(f^{-1}(s),g^{-1}(s,t);\mathbb{C})$ is q.u. of index $\leq n$.

For this purpose we investigate the Leray spectral sequence of g. Let $R^q g_*(\mathbb{C}_{\overline{X}})$ denote the sheaf on $\Delta\times\overline{\Delta}'$ which is associated to the presheaf $\Delta\times\overline{\Delta}' \supset U \to H^q(g^{-1}(U);\mathbb{C})$. Since g is C^∞-locally trivial over $(\Delta\times\overline{\Delta}')-D$, the restriction of $R^q g_*(\mathbb{C}_{\overline{X}})$ to $(\Delta\times\overline{\Delta}')-D$ is locally constant. If $z \in D-\{(0,0)\}$ then the singular points of $g^{-1}(z)$ are quadratic and their number is locally constant on $D-\{(0,0)\}$. It then follows from (4.2.iv) that the restriction of $R^q g_*(\mathbb{C}_{\overline{X}})$ to $D-\{(0,0)\}$ is also locally constant. If Δ'' is a complex disc centered at $0 \in \mathbb{C}$ and $\phi : \Delta'' \to \Delta\times\overline{\Delta}'$ is an analytic map with $\phi(0) = (0,0)$, $\phi(\Delta''^*) \subset (\Delta\times\overline{\Delta}')-D$, then the pull-back $\phi^*(\overline{X})$ is nonsingular (of dim n) outside $(0,x)$ and $\phi^*(g) : \phi^*(\overline{X}) \to \Delta''$ is a good representative of its germ at $(0,x)$. Our induction hypothesis says that the local system $\phi^* R^{n-1} g_*(\mathbb{C}_{\overline{X}})|\Delta''^*$ has q.u. monodromy of index $\leq n-1$.

Now fix a $t \in \partial\overline{\Delta}'$ and let $V \subset \Delta$ be open. Then the Leray spectral sequence of the mapping of pairs $g : (g^{-1}(V\times\overline{\Delta}'),g^{-1}(V\times\{t\})) \to (V\times\overline{\Delta}',V\times\{t\})$ is a natural spectral sequence whose E_2 term is

$$E_2^{pq} = H^p(V\times\overline{\Delta}',V\times\{t\};R^q g_*(\mathbb{C}_{\overline{X}}))$$

and which converges to $H^{p+q}(g^{-1}(V\times\overline{\Delta}'),g^{-1}(V\times\{t\});\mathbb{C})$, see Godement (1958) Th. 4.17.1, for the absolute case; the case of pairs is similar. Hence the theorem will follow from the basic

(5.15) *Proposition.* Let Δ and Δ' be copies of the complex unit disc, D an analytic curve in $\Delta\times\Delta'$, closed in $\Delta\times\overline{\Delta}'$ such that the projection $\pi : \Delta\times\overline{\Delta}' \to \Delta$ restricted to D is finite and unramified over Δ^*. Suppose we are given an $n \in \mathbb{N}$ and a sheaf $\mathcal{F}$ of complex vector spaces on

$E := \Delta^*\times\bar{\Delta}'$ such that $F|E-(D\cap E)$ and $F|(D\cap E)$ are locally constant and for any analytic map $\phi : \Delta'' \to \Delta\times\bar{\Delta}'$ (Δ'' a copy of the complex unit disc) with $\phi(0) \in D \cap (\{0\}\times\bar{\Delta}')$, $\phi(\Delta''^*) \subset E-(D\cap E)$, the local system $\phi^*F|\Delta''^*$ has q.u. monodromy of index $\leq n-1$. Then the sheaf F^q on Δ^* associated to $\Delta^* \supset V \to H^q(V\times\bar{\Delta}',V\times\{1\};\ F)$ is identically zero for $q \neq 1$ and F^1 is a local system with q.u. monodromy of index $\leq n$.

Proof. The hypotheses imply that any $s \in \Delta^*$ admits a neighbourhood V in Δ^* such that the quadruple $(V\times\bar{\Delta}',(V\times\bar{\Delta}') \cap D,\ V\times\{1\};F|V\times\bar{\Delta}')$ is V-isomorphic to the product of V with the restriction of this quadruple to $\{s\}\times\bar{\Delta}'$. This implies that F^q is a local system. Since $\{V\times\bar{\Delta}' : s \in V \text{ open}\}$ is a neighbourhood basis of $\{s\}\times\bar{\Delta}'$, it follows from a standard result (see for instance Godement (1958), Th. 4.11.1) that the stalk F^q_s can be identified with $H^q(\{s\}\times\bar{\Delta}',\{(s,1)\};F)$. Let $\Gamma \subset \{s\}\times\bar{\Delta}'$ be a tree whose ends are $(s,1)$ and the points of $D \cap (\{s\}\times\bar{\Delta}')$. Then there is a deformation retraction $r : \{s\}\times\bar{\Delta}' \to \Gamma$ so that the restriction map $H^q(\{s\}\times\bar{\Delta}',\{(s,1)\};F) \to H^q(\Gamma,\{(s,1)\};F)$ is an isomorphism (recall that $F|E-(D\cap E)$ is locally constant, so that $F|\{s\}\times\bar{\Delta}'$ may be identified with $r^*(F|\Gamma)$). Clearly $H^q(\Gamma,\{(s,1)\};F) = 0$ if $q \neq 1$. Hence $F^q = 0$ for $q \neq 1$.

It remains to prove that F^1 has q.u. monodromy of index $\leq n$. This we do with induction on the degree d of $\pi|D$. If $d = 0$, then F^1 is the trivial sheaf and there is nothing to prove. So suppose $d \geq 1$. Clearly, no generality is lost (and d remains the same) if we shrink Δ or even make a base change $s' \in \Delta \to s'^k \in \Delta$. Such a base change enables us to assume that the covering $\pi : D \cap E \to \Delta^*$ is trivial. We stick to this assumption throughout the following five steps which make up the rest of the proof.

Step 1. The proposition is true if D meets $\{0\}\times\Delta'$ in more than one point (so this assertion is vacuous if $d = 1$).

Proof. If this is the case, let T denote the set of $t \in \Delta'$ such that $(0,t) \in D$. For each $t \in T$ we choose a disc-like neighbourhood Δ'_t of t in Δ' such that the $\overline{\Delta}'_t$ are disjoint and do not meet $\partial\overline{\Delta}'$. By shrinking Δ we may assume that $D \subset \cup_t (\Delta\times\Delta'_t)$. Choose $p_t \in \partial\overline{\Delta}'_t$ and let $F^{1,t}$ denote the sheaf on Δ^* associated to $V \to H^1(V\times\overline{\Delta}'_t, V\times\{p_t\};F)$. Since $D \cap (\Delta\times\Delta'_t)$ has degree $< d$ over Δ, our induction hypothesis implies that $F^{1,t}$ is a local system with q.u. monodromy of index $\le n$. If Γ is a tree in $\overline{\Delta}'$ with $\Gamma \cap \partial\overline{\Delta}' = \{1\}$, $\Gamma \cap \partial\overline{\Delta}'_t = \{p_t\}$ $(t \in T)$, then for any $V \subset \Delta^*$ we have a chain of isomorphisms

$$\begin{array}{ccc} H^1(V\times(\overline{\Delta}',\{1\});F) & \overset{\cong}{\to} & H^1(V\times(\cup_t\overline{\Delta}'_t \cup \Gamma,\Gamma);F) \\ & & \Big\downarrow \cong \text{ excision} \\ & & H^1(V\times\cup_t(\overline{\Delta}'_t,\{p_t\});F) \overset{\cong}{\to} \oplus_t H^1(V\times(\overline{\Delta}'_t,\{p_t\});F). \end{array}$$

This establishes an isomorphism $F^1 \to \oplus_{t\in T} F^{1,t}$ and this proves that F^1 has q.u. monodromy of index $\le n$.

In view of step 1, it suffices to consider the case when D meets $\{0\}\times\Delta'$ in a single point, say in $\{(0,0)\}$. We make this assumption from now on. We further put $\partial E := \Delta^*\times\partial\overline{\Delta}'$ and we let G denote the local system on ∂E associated to $\partial E \supset U \to H^1(\pi^{-1}\pi(U),U;F)$. Notice, that if σ_0 denotes the section of $\pi : \partial E \to \Delta^*$ defined by $\sigma_0(s) = (s,1)$, then $F^1 \cong \sigma_0^*G$. So it is (more than) enough to show that G has q.u. monodromy of index $\le n$.

Step 2. There is a section σ of $\pi : \partial E \to \Delta^*$ such that σ^*G has q.u. monodromy of index $\le n$.

Proof. Since $D \cap E$ is the trivial covering over Δ^*, D decomposes into d irreducible components $D_1,\dots,D_d$ and each D_i is the graph of an analytic function $\phi_i : \Delta \to \Delta'$. Let m_i denote the order of vanishing of ϕ_i in $0 \in \Delta$ and put $m := \min_i m_i$. Then $m \geq 1$. Let c_i denote the coefficient of s^m in the Taylor expansion of ϕ_i in 0. A coordinate change of the form $(s,t) \to (s,t+\varepsilon s)$, ε small, enables us to assume that not all the c_i's are equal. (Although such a coordinate change need not preserve $\Delta\times\overline{\Delta}'$, it does preserve π and Δ and a little shrinking of $\overline{\Delta}'$ does not affect G.) Now choose $\rho > \max_i|c_i|$. Then after a possible shrinking of Δ, we have that $|\phi_i(s)| < \rho s^m$ for all i and $s \in \Delta$. Define $\psi : \Delta\times\overline{\Delta}' \to \Delta\times\overline{\Delta}'$ by $\psi(s,\tau) = (s,\rho s^m\tau)$. Then the strict transform of D under ψ meets $\{0\}\times\Delta'$ in $\{(0,\rho^{-1}c_i)\}_i$ and has therefore more than one element. So step 1 applies to ψ^*F and tells us that $(\psi^*F)^1$ is q.u. of index $\leq n$. Put

$$B = \{(s,t) \in \Delta^*\times\overline{\Delta}' : |t| \leq \rho|s|^m\}$$

$$\Sigma_0 = \{(s,\rho s^m) : s \in \Delta^*\}$$

$$\Sigma = \{(s,t) \in \Delta^*\times\overline{\Delta}' : t/|t| = (s/|s|)^m,\ \rho \leq |t| \leq 1\}$$

$$\Sigma_1 = \{(s,\sigma(s)) : s \in \Delta^*\} \text{ with } \sigma(s) = (s/|s|)^m.$$

Then for $V \subset \Delta^*$, we have natural isomorphisms

$$\begin{aligned}(\psi^*F)^1(V) &\cong H^1(B \cap \pi^{-1}(V),\ \Sigma_0 \cap \pi^{-1}(V);F)\\ &\cong H^1(\pi^{-1}(V),\ \Sigma \cap \pi^{-1}(V);F)\\ &\cong H^1(\pi^{-1}(V),\ \Sigma_1 \cap \pi^{-1}(V);F) \cong (\sigma^*G)(V),\end{aligned}$$

which proves that σ^*G is q.u. of index $\leq n$.

Step 3. $F|\partial E$ has q.u. monodromy of index $\leq n-1$.

Proof. The fundamental group of ∂E is the free abelian group on the two elements ξ,η defined by the factors $\Delta^*,\partial\overline{\Delta}'$. Let X_0 resp. Y_0 denote the corresponding automorphisms of $F|\partial E$. If m is as in the previous step, and

$p,q \in \mathbb{N}$ are such that $q/p < m$, then $\phi_{p,q} : \Delta'' \to \Delta\times\Delta'$ (Δ'' a small disc about $0 \in \mathbb{C}$), $\phi_{p,q}(\tau) = (\tau^p,\tau^q)$ has the property that $\phi_{p,q}^{-1}(D)$ contains $0 \in \Delta''$ as an isolated point. After shrinking Δ'', Δ''^* will then map in $E-(D \cap E)$ and the homotopy $(u,\tau) \in [0,1]\times\Delta''^* \to (\tau^p,(1-u)\tau^q + u(\tau/|\tau|)^q)$ will take place in $E-(D \cap E)$, showing that $\phi_{p,q}|\Delta''^*$ represents the free homotopy class $\xi^p\eta^q$. By assumption $X_0^pY_0^q$ is then q.u. of index $\leq n-1$. So if $k \in \mathbb{N}$ is such that X_0^k and Y_0^k are unipotent, then the semi-group generated by $X_0^{3k}Y_0^k$, $X_0^{2k}Y_0^k$ is unipotent of index $\leq n-1$. By lemma (5.13) the same is then true for the group generated by these elements. As this group consists of all the kth powers of monomials in X_0 and Y_0, step 3 follows.

Step 4. Let ∂G denote the local system on ∂E associated to

$$\partial E \supset U \to H^1(\pi^{-1}\pi(U) \cap \partial E,U;F).$$

Then there is a natural isomorphism $\alpha : \partial G \to F|\partial E$. If $\gamma : G \to F|\partial E$ denotes the composite of the restriction map $\beta : G \to \partial G$ with α and $\delta : F|\partial E \to G$ is the usual coboundary, then $1 + \gamma\delta : F|\partial E \to F|\partial E$ is just the automorphism Y_0 of $F|\partial E$ induced by the positive generator η of $\pi(\partial\overline{\Delta}')$. Moreover, the automorphism Y of G induced by η acts as the identity on $\mathrm{Ker}(\gamma)$ and $\mathrm{Coker}(\delta)$.

Remark. Presumably, $Y = 1 + \delta\gamma$.

Proof. If $(s,t) \in \partial E$, then $G_{(s,t)} \cong H^1(\{s\}\times(\overline{\Delta}',\{t\});F)$ and $\partial G_{(s,t)} = H^1(\{s\}\times(\partial\overline{\Delta}',\{t\});F)$. Define a relative homeomorphism $\psi : ([0,1],\{0,1\}) \to \{s\}\times(\partial\overline{\Delta}',\{t\})$ by $\psi(\tau) = (s,e^{2\pi i\tau}t)$. Then ψ induces an isomorphism $\psi^* : H^1(\{s\}\times(\partial\overline{\Delta}',\{t\});F) \to H^1([0,1],\{0,1\};\psi^*F)$. Since the local system ψ^*F is constant, there is a unique way to identify it with the constant sheaf $F_{(s,t)}$ such that in $\{0\}$ this is the natural identification. The cup product then induces an isomorphism

$$\mu : H^1([0,1],\{0,1\};\mathbf{Z})\otimes_{\mathbf{Z}} F_{(s,t)} \to H^1([0,1],\{0,1\};\psi^*F).$$

Picking the canonical generator u of $H^1([0,1],\{0,1\};\mathbf{Z})$ identifies the left hand side with $F_{(s,t)}$ and so $\alpha := \mu^{-1}\circ\psi^*$ establishes an isomorphism between ∂G and $F|\partial E$.

Nex we consider the commutative diagram

$$\begin{array}{ccccc}
F_{(s,t)} & \xrightarrow{\delta} H^1(\{s\}\times(\bar{\Delta}',\{t\});F) \xrightarrow{\beta} & H^1(\{s\}\times(\partial\bar{\Delta}',\{t\});F) & & \\
\downarrow \psi_0^* & & \downarrow \psi^* \cong & \searrow \alpha \cong & \\
H^0(\{0,1\};\psi^*F) & \xrightarrow{\delta'} & H^1([0,1],\{0,1\};\psi^*F) & & \\
\cong \uparrow \mu_0 & & \mu \uparrow \cong & & \\
H^0(\{0,1\};\mathbf{Z})\otimes_{\mathbf{Z}} F_{(s,t)} & \xrightarrow{\delta''\otimes 1} & H^1([0,1],\{0,1\};\mathbf{Z})\otimes_{\mathbf{Z}} F_{(s,t)} & \cong F_{(s,t)} &
\end{array}$$

where $\psi_0 = \psi|\{0,1\}$ and μ_0 is defined analogously to μ. So $\mu_0^{-1}\psi_0^*(x) = \{0\}\otimes x + \{1\}\otimes Y(x)$. Since $\delta''(\{0\}) = -u$, $\delta''(\{1\}) = u$, it follows that $\gamma\delta(x) = \alpha\beta\delta(x) = (\delta''\otimes 1)(\mu_0^{-1}\circ\psi_0^*)(x) = Y(x)-x$.

Now look at the following pieces of exact sequences in which β and δ occur:

$$H^1(\{s\}\times(\bar{\Delta}',\partial\bar{\Delta}');F) \to H^1(\{s\}\times(\Delta',\{t\});F) \xrightarrow{\beta} ..$$

$$.. \xrightarrow{\delta} H^1(\{s\}\times(\partial\bar{\Delta}',\{t\});F) \to H^1(\{s\}\times\Delta';F).$$

Since η acts as the identity on the first resp. the last term, it follows that $\mathrm{Ker}(Y-1) \subset \mathrm{Ker}(\beta) = \mathrm{Ker}(\gamma)$ and $\mathrm{Im}(Y-1) \subset \mathrm{Im}(\delta)$.

To finish the proof we need

(5.16) *Lemma*. Let

$$V_0 \overset{\delta}{\to} V \overset{\gamma}{\to} V_0$$

be a diagram of $\mathbb{C}$-vector spaces and homomorphisms and let $(Y_0,Y) \in \mathrm{Aut}(V_0)\times\mathrm{Aut}(V)$ be an automorphism of the diagram (so $\delta Y_0 = \gamma Y = Y_0\gamma$) with $Y_0 = 1 + \gamma\delta$ and assume that Y induces the identity on $\mathrm{Ker}(\gamma)$ and $\mathrm{Coker}(\delta)$. Let $(X_0,X) \in \mathrm{Aut}(V_0)\times\mathrm{Aut}(V)$ be an automorphism of the diagram which commutes with (Y_0,Y) such that X_0,Y_0 generate a group of q.u. automorphisms of index $\leq n-1$. If X is q.u. of index $\leq n$, then so is the group generated by X and Y.

Proof. The assumption that Y acts trivially on $\mathrm{Ker}(\gamma)$ and $\mathrm{Coker}(\delta)$ means that Y is of the form $1 + \delta Y_1\gamma$ for some $Y_1 \in \mathrm{End}(V_0)$. Since $Y\delta = \delta Y_0$, we have $\delta Y_1\gamma\delta = \delta\gamma\delta$ and similarly $\gamma Y = Y_0\gamma$ implies $\gamma\delta Y_1\gamma = \gamma\delta\gamma$. Now let $k \in \mathbb{N}$ be such that X^k, X_0^k and Y_0^k are all unipotent and put $\delta' := \Sigma_{i=1}^{k}\binom{k}{i}(\delta\gamma)^{i-1}\delta$. Then

$$\gamma\delta' = \Sigma_{i=1}^{k}\binom{k}{i}(\gamma\delta)^{i} = Y_0^k-1$$

$$\delta' Y_1\gamma = \Sigma_{i=1}^{k}\binom{k}{i}(\delta\gamma)^{i-1}\delta Y_1\gamma = \Sigma_{i=1}^{k}\binom{k}{i}(\delta Y_1\gamma)^{i} = Y^k-1$$

$$\delta' Y_1\gamma\delta' = \delta'\gamma\delta', \quad \gamma\delta' Y_1\gamma = \gamma\delta'\gamma.$$

Put $N := X^k-1$ and $N_0 := X_0^k-1$. By lemma (5.13) we only need to show that $(X^k-1)^p(Y^k-1)^q = N^p(\delta' Y_1\gamma)$ is zero for $p+q = n+1$. By the same lemma we are given that $N_0^p(\gamma\delta')^q = 0$ for $p+q = n$. So if $q \geq 1$, then $N^{n-q}(\delta' Y_1\gamma)^{q+1} = N^{n-q}(\delta'\gamma)^{q+1} = \delta' N_0^{n-q}(\gamma\delta')^q\gamma = 0$. Furthermore, $N^n(\delta' Y_1\gamma) = \delta' N_0^n Y_1\gamma = 0$ and $N^{n+1} = 0$, by assumption.

Step 5. G has q.u. monodromy of index $\leq n$.

Proof. Let $s \in \Delta^*$. We apply the lemma with $V_0 := F_{\sigma(s)}$, $V := G_{\sigma(s)}$, β, γ, Y_0, Y as in step 4 and X_0 resp. X the automorphisms of $F_{\sigma(s)}$ resp. $G_{\sigma(s)}$ induced by the action of the image of the canonical generator of

$\pi(\Delta^*,s)$ under σ_*. Then it follows from the previous steps that the conditions of the lemma are satisfied so that X and Y generate a q.u. group of automorphisms of index $\leq n$. Since this group is the whole monodromy group of the local system G, the assertion follows.

6 DEFORMATIONS

In previous sections we studied map-germs $f : (\mathbb{C}^{n+k},0) \to (\mathbb{C}^k,0)$ defining an icis $(X_0,0)$ of dim n without worrying to what extent invariants of f were really invariants of $(X_0,0)$. In order to answer this question we must fix $(X_0,0)$ and compare the various $f : (\mathbb{C}^{n+k},0) \to (\mathbb{C}^k,0)$ which define $(X_0,0)$ (so k varies also). A simple method by which we get new defining equations out of old is to choose a germ $g : (\mathbb{C}^{n+k}\times\mathbb{C}^{\ell},0) \to (\mathbb{C}^k,0)$ with $g(z,0) = f(z)$: then $F : (\mathbb{C}^{n+k+\ell},0) \to (\mathbb{C}^{k+\ell},0)$, $F(z,u) = (g(z,u),u)$ defines the same singularity as f. It appears that there exist germs $f : (\mathbb{C}^{n+k},0) \to (\mathbb{C}^k,0)$ defining $(X_0,0)$ which are in some sense saturated with respect to this process: any extension F of f obtained as above is analytically equivalent to the trivial extension $f\times 1_{(\mathbb{C}^{\ell},0)}$. We call them versal deformations of $(X_0,0)$ (allthough this is not quite the way by which we shall define this notion). Versal deformations appear to be unique in the sense that of two versal deformations of $(X_0,0)$ one is always analytically equivalent to a trivial extension of the other. These (and other) properties are proven in (6.C). Of the two sections preceding it, the first one is concerned with (relative) differentials in the analytic category. This notion is needed here in order to develop the Kodaira-Spencer map in an adequate setting (in B); it will reappear in Ch.8 when we deal with (relative) De Rham cohomology. In the last section (D) we discuss some analytic properties of versal deformations.

6.A *Relative differentials*

We begin with a provisional definition of differentials which is useful in doing computations. Let $(X,\mathcal{O}_X)$ be an analytic space in an open $U \subset \mathbb{C}^N$, defined by an ideal $I_X = (g_1,\ldots,g_\ell)\mathcal{O}_U$. Then we let Ω_X denote the restriction of

$$\mathcal{O}_U.\{dz_1,\ldots,dz_N\}/(\mathcal{O}_U.\{dg_1,\ldots,dg_\ell\}+I_X.\{dz_1,\ldots,dz_N\})$$

to X. It is clear that this is a coherent $\mathcal{O}_X$-module and that the map $f \in \mathcal{O}_U \to df$ induces a $\mathbb{C}$-linear map $d_X : \mathcal{O}_X \to \Omega_X$ satisfying $d_X(\phi\psi) = \phi d_X(\psi) + \psi d_X(\phi)$. In case X is reduced and nonsingular, Ω_X can be identified, as one may expect, with the sheaf of holomorphic sections of the cotangent bundle of X. What is not so clear yet is that Ω_X is invariantly defined. There are at least two invariant characterizations of Ω_X: one is an explicit (invariant) construction of Ω_X on the diagonal of X×X (compare Teissier (1976)), another regards Ω_X as the solution of a universal problem (Grauert & Remmert (1971)). We outline the last approach.

Fix a ring k and a commutative k-algebra A. A k-*derivation of* A *into an* A-*module* M is simply a k-linear map $D : A \to M$ satisfying the Leibniz rule: $D(a_1a_2) = a_1Da_2 + a_2Da_1$. It can be shown that there is a universal k-derivation $d : A \to \Omega_{A/k}$: it has the property that any k-derivation of A into an A-module M is the composite of d and a *unique* A-homomorphism $\Omega_{A/k} \to M$, see e.g. Matsumura (1980). It is clear that $\Omega_{A/k}$ is then unique up to canonical isomorphism. When A is a finitely generated k-algebra, $\Omega_{A/k}$ will be a finitely generated A-module, but in our case (e.g. $k = \mathbb{C}$, $A = \mathcal{O}_{X,x}$), A is usually not finitely generated and $\Omega_{A/k}$ will be too large. We do get the right object however, if we restrict ourselves to $\mathcal{O}_{X,x}$-modules M which are separated for the $m_{X,x}$-adic topology (i.e. $\bigcap_{k=1}^{\infty} m_{X,x}^{k} M = \{0\}$): for such modules, $d_X : \mathcal{O}_{X,x} \to \Omega_{X,x}$ is indeed the universal derivation. So the stalks $\Omega_{X,x}$ are invariantly characterized and hence so is Ω_X.

If $f^* : \mathcal{O}_{S,s} \to \mathcal{O}_{X,x}$ is a homomorphism of local analytic $\mathbb{C}$-algebras, then composing f^* with d_X gives a $\mathbb{C}$-derivation of $\mathcal{O}_{S,s}$ into $\Omega_{X,x}$. Now, $\Omega_{X,x}$ is a separated $\mathcal{O}_{S,s}$-module (because $\mathcal{O}_{X,x}$ is such and $\Omega_{X,x}$ is a finitely generated $\mathcal{O}_{X,x}$-module), so that $d_{X,x} \circ f^*$ factors over a unique $\mathcal{O}_{S,s}$-homomorphism

$$df : \Omega_{S,s} \to \Omega_{X,x}.$$

We let $\Omega_{f,x}$ denote the quotient module of $\Omega_{X,x}$ by the $\mathcal{O}_{X,x}$-submodule generated by the image of df. Notice that the composite map $d_f : \mathcal{O}_{X,x} \to \Omega_{f,x}$ is now an $\mathcal{O}_{S,s}$-derivation. In fact, it can be shown that it is the universal $\mathcal{O}_{S,s}$-derivation of the $\mathcal{O}_{S,s}$-algebra $\mathcal{O}_{X,x}$ with respect to separated $\mathcal{O}_{X,x}$-modules.

Thus we find that for any analytic space $(X,\mathcal{O}_X)$ we have the coherent $\mathcal{O}_X$-module of *(absolute) differentials* Ω_X. A morphism $f : (X,\mathcal{O}_X) \to (S,\mathcal{O}_S)$ of analytic spaces induces an $\mathcal{O}_S$-homomorphism $df : \Omega_S \to \Omega_X$ and the cokernel of the induced $\mathcal{O}_X$-homomorphism $\mathcal{O}_X \otimes_{\mathcal{O}_S} \Omega_S \to \Omega_X$ is the coherent $\mathcal{O}_X$-module of *relative differentials* Ω_f which comes equipped with an $\mathcal{O}_S$-derivation $d_f : \mathcal{O}_X \to \Omega_f$. An argument analogous to the one employed in (4.3) shows that Ω_f behaves well with respect to base change: if we are given a cartesian square

$$\begin{array}{ccc} X' & \xrightarrow{g'} & X \\ {\scriptstyle f'}\downarrow & & \downarrow{\scriptstyle f} \\ S' & \xrightarrow[g]{} & S \end{array}$$

then the natural map $\mathcal{O}_{X'} \otimes_{\mathcal{O}_X} \Omega_f \to \Omega_{f'}$ is an isomorphism of $\mathcal{O}_{X'}$-modules. In particular we have (take for S' a point $s \in S$) $\Omega_f/m_{S,s}\Omega_f \cong \Omega_{X_s}$. So Ω_f may be thought of as the sheaf of differentials along the fibres of f. Notice by the way, that for the construction of Ω_f the target S of f is rather irrelevant: it is the fibres of f that count. The critical locus of a morphism $f : X \to S$, all of whose fibres have pure dimension n, can be conveniently expressed with the help of relative differentials: C_f is just the nth Fitting ideal of Ω_f (proof is left to the reader).

By construction, $\mathrm{Hom}_{\mathcal{O}_X}(\Omega_X, \mathcal{O}_X)$ is the sheaf of all $\mathbb{C}$-derivations of $\mathcal{O}_X$ into $\mathcal{O}_X$; we shall denote it by θ_X. Similarly, we denote $\mathrm{Hom}_{\mathcal{O}_X}(\Omega_f, \mathcal{O}_X)$ by θ_f; it is the sheaf of holomorphic vector fields along the fibres of f. It is a coherent $\mathcal{O}_X$-module, of course, but it does not, in general, commute with base change. In chapter 9 we will prove a result (lemma (9.5)) which produces examples of this phenomenon in a systematic fashion.

6.B *The Kodaira-Spencer map*

(6.1) Let $f : (\mathbb{C}^{n+k}, x) \to (\mathbb{C}^k, 0)$ define an icis of dim n. For reasons of notation (and also to emphasize the intrinsic character of the subsequent constructions) it will be convenient to write $X = \mathbb{C}^{n+k}$ and $S = \mathbb{C}^k$. We put

$$\theta(f)_x := \mathcal{O}_{X,x} \otimes_{\mathcal{O}_{S,0}} \theta_{S,0}.$$

This is a free $\mathcal{O}_{X,x}$-module of rank k (a basis is given by $\frac{\partial}{\partial t_1}, \ldots, \frac{\partial}{\partial t_k}$). One way to give an intrinsic meaning to the derivative of f is by viewing it as an $\mathcal{O}_{X,x}$-homomorphism $\partial f : \theta_{X,x} \to \theta(f)_x$, $\frac{\partial}{\partial z_\nu} \to \Sigma_{\kappa=1}^{k} \frac{\partial f_\kappa}{\partial z_\nu} \frac{\partial}{\partial t_\kappa}$;

this is simply the $\mathcal{O}_{X,x}$-dual of the natural map $\mathcal{O}_{X,x}\otimes_{\mathcal{O}_{S,0}}\Omega_{S,0} \to \Omega_{X,x}$. It is customary to denote the cokernel of ∂f by $T^1_{f,x}$. We omit the upper index and write $T_{f,x}$ instead. Since $C_f \subset \mathcal{O}_{X,x}$ is the ideal generated by the determinants of the k×k minors of $(\frac{\partial f_\kappa}{\partial z_\nu})$, it follows from Cramer's rule that $C_f\theta(f)$ is contained in the image of ∂f. So $T_{f,x}$ is in a natural way an $\mathcal{O}_{C,x}$-module. Hence $T_{f,x}$ is a finitely generated $\mathcal{O}_{D,0}$-module (because $\mathcal{O}_{C,x}$ is). In particular, $T_{f,x}/m_{S,0}T_{f,x}$ (which we shall denote by $T_{X_0,x}$) is of finite $\mathbb{C}$-dimension. As the notation suggests $T_{X_0,x}$ only depends on (X_0,x). This can be verified by a straightforward, though somewhat tedious computation. In case n > 0 this will also follow from the discussion below.

Consider the sequence of $\mathcal{O}_{X,x}$-modules

$$S : 0 \to \mathcal{O}_{X,x}\otimes_{\mathcal{O}_{S,0}}\Omega_{S,0} \xrightarrow{1\otimes df} \Omega_{X,x} \to \Omega_{f,x} \to 0$$

which we claim to be exact. For this we need only verify that $1\otimes df$ is injective. At points of X where f is a submersion this is clear. As such points are dense in (X,x), $1\otimes df$ is injective in x, too. If we view S as a resolution of $\Omega_{f,x}$ by free $\mathcal{O}_{X,x}$-modules, then we see that

$$T_{f,x} = \mathrm{Ext}^1_{\mathcal{O}_{X,x}}(\Omega_{f,x},\mathcal{O}_{X,x})$$

and that the higher Ext's vanish. If we reduce S modulo $m_{X,x}$ we get

$$S(x) : 0 \to \mathcal{O}_{X_0,x}\otimes_{\mathbb{C}}T_0(S)^* \to \mathcal{O}_{X_0,x}\otimes_{\mathcal{O}_{X,x}}\Omega_{X,x} \to \Omega_{X_0,x} \to 0,$$

where $T_0(S)$ denotes the tangent space of S in 0. For n > 0 the same argument used to prove the exactness of S proves the exactness of $S(x)$. So

$$T_{X_0,x} = \mathrm{Ext}^1_{\mathcal{O}_{X_0,x}}(\Omega_{X_0,x},\mathcal{O}_{X_0,x})$$

in this case (and the higher Ext's vanish). The $\mathbb{C}$-dimension of $T_{X_0,x}$, is called the Tjurina number, after Tjurina who was among the first to investigate deformations of isolated singularities. We shall denote this number by $\tau(X_0,x)$.

Example 1. If $f : (\mathbb{C}^{n+1},x) \to (\mathbb{C},0)$ defines an isolated singularity (X_0,x) then $T_{X_0,x} \cong \mathcal{O}_{\mathbb{C}^{n+1},x}/(\frac{\partial f}{\partial z_1},\ldots,\frac{\partial f}{\partial z_{n+1}},f)\mathcal{O}_{\mathbb{C}^{n+1},x}$. Comparing this with the formula for the Milnor number, we see that $\tau(X_0,x) \le \mu(X_0,x)$. It was proven recently that this inequality holds for any icis of dim > 0, (Looijenga & Steenbrink (1983)). We return to this question in (9.A).

The natural map $\theta_{S,0} \to \theta(f)_x$ can be composed with the surjection $\theta(f)_x \to T_{f,x}$ to give the *Kodaira-Spencer map*

$$\rho_{f,x} : \theta_{S,0} \to T_{f,x}$$

This is an $\mathcal{O}_{S,0}$-homomorphism. If we reduce mod $m_{X_0,x}$ we get just a $\mathbb{C}$-linear map (the *reduced Kodaira-Spencer map*):

$$\rho_f(x) : T_0(S) \to T_{X_0,x}$$

By Nakayama's lemma, $\rho_{f,x}$ is surjective if and only if $\rho_f(x)$ is.

(6.2) *Proposition*. $T_{f,x}$ and $\rho_{f,x}$ behave well with respect to base change: if we have a cartesian diagram of non-singular germs

$$\begin{array}{ccc} (X',x') & \xrightarrow{\tilde{g}} & (X,x) \\ {\scriptstyle f'}\downarrow & & \downarrow{\scriptstyle f} \\ (S',0) & \xrightarrow{g} & (S,0) \end{array}$$

in which f (and hence f') defines an icis, then $T_{f',x'}$ can be naturally identified with $\mathcal{O}_{X',x'}\otimes_{\mathcal{O}_{X,x}} T_{f,x}$ as $\mathcal{O}_{X',x'}$-modules (and hence as $\mathcal{O}_{S',0}$-modules) and via this identification, $\rho_{f',x'}$ is the composite

$$\theta_{S',0} \xrightarrow{\partial g} \mathcal{O}_{S',0}\otimes_{\mathcal{O}_{S,0}}\theta_{S,0} \xrightarrow{1\otimes\rho_{f,x}} \mathcal{O}_{S',0}\otimes_{\mathcal{O}_{S,0}} T_{f,x} \cong \mathcal{O}_{X',x'}\otimes_{\mathcal{O}_{X,x}} T_{f,x}$$

where the first map is the $\mathcal{O}_{S',0}$-dual of $dg : \Omega_{S,0} \to \Omega_{S',0}$.

Proof. From the commutative diagram of $\mathcal{O}_{X,x}$-modules

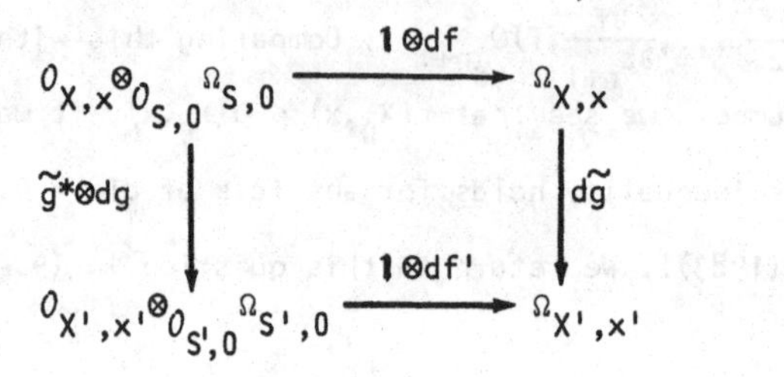

we derive a natural $\mathcal{O}_{X,x}$-homomorphism from $T_{f,x}$ to $T_{f',x'}$. This induces an $\mathcal{O}_{S',0}$-homomorphism $\phi : \mathcal{O}_{S',0}\otimes_{\mathcal{O}_{S,0}} T_{f,x} \to T_{f',x'}$, which fits in a commutative diagram:

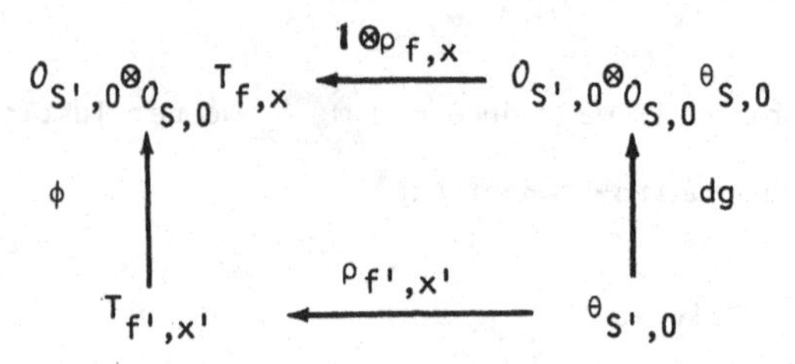

For any $\mathcal{O}_{X,x}$-module M, we may identify $\mathcal{O}_{S',0}\otimes_{\mathcal{O}_{S,0}} M$ with $\mathcal{O}_{X',x'}\otimes_{\mathcal{O}_{X,x}} M$ (by assumption this is so for $M = \mathcal{O}_{X,x}$) and thus ϕ may be viewed as an $\mathcal{O}_{X',x'}$-homomorphism. It remains to show that ϕ is an isomorphism. For this purpose it is convenient to write $(\tilde{g},g)$ as the composite of an embedding (the graph of $(\tilde{g},g)$) and a projection (the projection of $f\times 1_{S'} : (X\times S',x\times 0) \to (S\times S',0\times 0)$ onto f). In the projection case it is clear that T behaves well with base change, so we only need to consider the case when $(\tilde{g},g)$ is an embedding. The assumption that (X',x') is nonsingular implies that g is transversal to f. This means that we can find a coordinate system $t_1,\dots,t_k$ for $(S,0)$ such that $(S',0)$ is defined by $t_1=\dots=t_{k'}=0$ and $f_1,\dots,f_{k'}$ extend to a coordinate system $z_1=f_1,\dots,z_{k'}=f_{k'},z_{k'+1},\dots,z_{n+k}$ of (X,x). Then the matrix of ∂f has the form

$$\left[\begin{array}{c|c} I & 0 \\ \hline * & \left(\frac{\partial f_\kappa}{\partial z_\nu}\right) \end{array}\right]$$

The reduction of the matrix is in the lower right hand corner modulo the ideal $I_{S',0}\mathcal{O}_{X,x}=\{z_1,\dots,z_{k'}\}\mathcal{O}_{X,x}$ is just the matrix of $\partial f'$. It easily follows that $T_{f',x}$ is the reduction of $T_{f,x}$ modulo $I_{S',0}\mathcal{O}_{X,x}$.

The following proposition is the central result of this chapter. It gives us a clue about the rôle the Kodaira-Spencer map will play.

(6.3) *Proposition*. Let $f^0,f^1:(\mathbb{C}^{n+k},x)\to(\mathbb{C}^k,0)$ both define the same icis $(X_0,x)\subset(\mathbb{C}^{n+k},x)$. If $\rho_{f^0,x}$ and $\rho_{f^1,x}$ are both surjective, then there exist analytic automorphisms $h:(\mathbb{C}^{n+k},x)\circlearrowleft$ and $h':(\mathbb{C}^k,x)\circlearrowleft$ such that $h'\circ f^1=f^0\circ h$ and h is the identity on (X_0,x).

Proof. We prove this proposition in four steps. Put $f^u(z)=(1-u)f^0(z)+uf^1(z)$ and define $F:(\mathbb{C}^{n+k},x)\times\mathbb{C}\to(\mathbb{C}^k,0)\times\mathbb{C}$ by $F(z,u)=(f(z,u),u)$. As before we write X for $\mathbb{C}^{n+k}$ and S for $\mathbb{C}^k$.

Step 1. The set U of $u\in\mathbb{C}$ for which f^u defines (X_0,x) and ρ_{f^u} is surjective in the complement of a finite subset of $\mathbb{C}$.

If we write $f^1_\kappa=\Sigma^k_{\lambda=1}\phi_{\kappa\lambda}f^0_\lambda$, $\phi_{\kappa\lambda}\in\mathcal{O}_{X,x}$, then $f^u_\kappa=\Sigma^k_{\lambda=1}((1-u)\delta_{\lambda\kappa}+u\phi_{\kappa\lambda})f^0_\lambda$. So $f^u_1,\dots,f^u_k$ generate the same ideal in $\mathcal{O}_{X,x}$

if the matrix $((1-u)\delta_{\lambda\kappa}+u\phi_{\kappa\lambda})$ is invertible. This is the case if u is not a root of the polynomial $\det((1-u)\delta_{\lambda\kappa}+u\phi_{\kappa\lambda}(0))$, which excludes finitely many u. For any such u, we can form $\rho_{f^u}(x) \in \mathrm{Hom}_{\mathbb{C}}(T_0(S),T_{X_0,x})$. Since $\rho_{f^u}(x) = (1-u)\rho_{f^0}(x) + u\rho_{f^1}(x)$ and $\rho_{f^0}(x)$ is surjective, it follows that the set of u for which ρ_{f^u} fails to be surjective is a proper algebraic subset of $\mathbb{C}$ and hence finite.

Step 2. Suppose that for any $u \in U$ we can find vector field germs $v \in \theta_{X\times U,x\times u}$ and $w \in \theta_{S\times U,0\times u}$ of the form

$$v = v_1\frac{\partial}{\partial z_1}+\ldots+v_{n+k}\frac{\partial}{\partial z_{n+k}}+\frac{\partial}{\partial u} \text{ with } v_\nu \in m_{S,0}\mathcal{O}_{X\times U,x\times u}$$

$$w = w_1\frac{\partial}{\partial t_1}+\ldots+w_k\frac{\partial}{\partial t_k}+\frac{\partial}{\partial u} \text{ with } w_k \in m_{S,0}\mathcal{O}_{S\times U,0\times u}$$

such that $dF(v) = w\circ F$. Then the proposition follows.

Let $u^0 \in U$. Since $v = \frac{\partial}{\partial u}$ on the u-axis, integration of v gives a neighbourhood U' of u^0 in U and an analytic isomorphism $H : (X,x)\times U' \circlearrowleft$ defined by $H(z,u) = \gamma_z(u-u^0)$, where γ_z denotes the integral curve of v with $\gamma_z(0) = (z,u^0)$. Because v projects to $\frac{\partial}{\partial u}$, H is of the form $H(z,u) = (h^u(z),u)$ and the assumption that $v_\nu \in m_{S,0}\mathcal{O}_{X\times U,x\times u}$ implies that H is the identity on $(X_0,x)\times U'$. Similarly, integration of w defines an isomorphism $H' : (S,0)\times U' \circlearrowleft$ of the form $H'(t,u) = (h'^u(t),u)$ with $h'(t,u^0) = t$ and $h'(0,u) = 0$. The condition $dF(v) = w\circ F$ implies that F maps integral curves of v onto integral curves of w, so that we have a commutative diagram

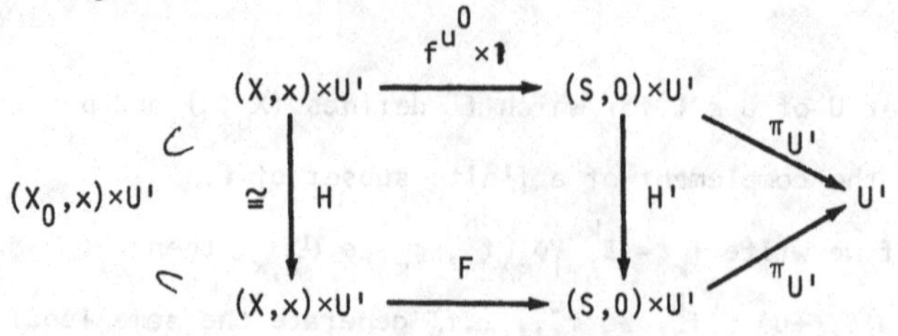

$$\begin{array}{ccccccc}
 & & (X,x)\times U' & \xrightarrow{f^{u^0}\times 1} & (S,0)\times U' & & \\
 & \nearrow\subset & \cong\downarrow H & & \downarrow H' & \searrow \pi_{U'} & \\
(X_0,x)\times U' & & & & & & U' \\
 & \searrow\subset & & & & \nearrow \pi_{U'} & \\
 & & (X,x)\times U' & \xrightarrow{F} & (S,0)\times U' & &
\end{array}$$

This shows that for any $u \in U'$ we have $f^u \circ h^u = h'^u \circ f^u$ where h^u and h'^u are isomorphisms and h^u is the identity on (X_0, x). The proposition now follows from the connectedness of U.

Step 3. Let $u \in U$ and denote by $\theta'_{X\times U, x\times u}$, $\theta'_{S\times U, 0\times u}$, $\theta(F)'_{x\times u}$ the set of elements of resp. $\theta_{X\times U, x\times u}$, $\theta_{S\times U, 0\times u}$, $\theta(F)_{x\times u}$ annihilated by resp. $\partial(\pi_U \circ F)$, $\partial\pi_U$, $\partial\pi_U$ (i.e. having no $\frac{\partial}{\partial u}$ component). Then

$$\theta(F)'_{x\times u} = \partial F(\theta'_{X\times U, x\times u}) + \theta'_{S\times U, 0\times u}.$$

The inclusion $\supset$ is clear. Let then M denote the left hand side modulo the right hand side. This is in a natural way an $\mathcal{O}_{S\times U, 0\times u}$-module. Now $M/m_{S\times U, 0\times u}M$ can be identified with

$$\theta(f^u)_x / \partial f^u(\theta_{X,x}) + \theta_{S,0} + m_{S,0}\theta(f^u)_x.$$

This is just Coker $(\rho_{f^u}(x))$ and hence trivial. It follows from Malgrange's form of the preparation theorem (Narasimhan (1966)) that then $M = 0$. So left hand side and right hand side are equal.

Step 4. Proof that v and w as in step 2 exist.

Let $v' \in \theta_{X\times U, x\times u}$ resp. $w' \in \theta_{S\times U, 0\times u}$ be given by $\frac{\partial}{\partial u}$. Then

$$\partial F(v') - w' = \Sigma_{\kappa=1}^{k} (f_\kappa^1 - f_\kappa^0) \frac{\partial}{\partial t_\kappa}$$

may be regarded as an element of $m_{S,0}\theta(F)'_{x\times u}$.
Multiplying the equation of step 3 with $m_{S,0}$ we find that there exist $v'' \in m_{S,0}\theta'_{X\times U, x\times u}$ and $w'' \in m_{S,0}\theta'_{S\times U, 0\times u}$ such that $\partial F(v') - w' = \partial F(v'') + w''$.
Then $v := v' - v''$ and $w := w' + w''$ are as required.

6.C *Versal deformations*

(6.4) For what is going to follow it will be convenient to have at our disposal the notion of the *deformation category* of a fixed icis (X_0,x_0) of dim n. A *deformation* of (X_0,x_0) will be simply a realization of (X_0,x_0) as the fibre of a map-germ $(\mathbb{C}^{n+k},0) \to (\mathbb{C}^k,0)$; in more invariant terms: it will consist of a map-germ $f : (X,x) \to (S,s)$ between non-singular germs with $\dim(X,x)-\dim(S,s) = n$ and an isomorphism ι of (X_0,x_0) onto the fibre (X_s,x) of f (so ι should induce an isomorphism between $\mathcal{O}_{X_s,x}/m_{S,s}\mathcal{O}_{X_s,x}$ and $\mathcal{O}_{X_0,x_0}$). A *morphism* from a deformation (ι',f') to another (ι,f) is a pair of map-germs $(\tilde{g},g)$ such that the diagram

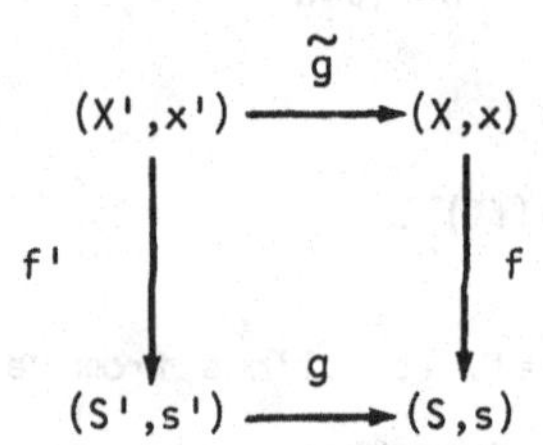

is cartesian and $\tilde{g}\circ\iota' = \iota$. A deformation (ι,f) of (X_0,x_0) is called *versal* (some authors say: *complete*) if for any deformation (ι',f') of (X_0,x_0) there exists a morphism $(\tilde{g},g)$ from (ι',f') to (ι,f). Notice that we do not require this morphism to be unique in any sense. If however, the derivative of g in s', $\partial g(s') : T_{s'}(S') \to T_s(S)$ is unique, then we say that (ι,f) is *miniversal* (or *semi-universal*). The perhaps more natural requirement that g be unique appears to be too strong: such deformations do not exist if (X_0,x_0) is singular.

(6.5) *Theorem*. Let (X_0,x_0) be an icis. Then

(i) A deformation of (X_0,x_0) is versal if and only if its Kodaira-Spencer map is surjective.

(ii) Two versal deformations of (X_0,x_0) are isomorphic if their base germs have the same dimension.

(iii) (X_0,x_0) admits a miniversal deformation and any two are isomorphic.

(iv) A deformation of (X_0,x_0) is miniversal if and only if its reduced Kodaira-Spencer map is an isomorphism.

We begin the proof with a lemma.

(6.6) *Lemma.* There exists a deformation of (X_0,x_0) whose reduced Kodaira-Spencer map is an isomorphism. Any deformation of (X_0,x_0) admits a morphism to a deformation with surjective Kodaira-Spencer map.

Proof. As for the first part, let $(\iota, f : (X,x) \to (\mathbb{C}^k,0))$ be a deformation of (X_0,x_0) with $\partial f(x) = 0$ (such deformations exist: if $f' : (X',x') \to (\mathbb{C}^{k'},0)$ is any deformation of (X_0,x_0), then after a coordinate change in $(\mathbb{C}^{k'},0)$ we may assume that $df'_1(x) = \ldots = df'_k(x) = 0$, $df'_{k+1}(x),\ldots,df'_{k'}(x)$ are linearly independent. Let f be the restriction of f' to $(\mathbb{C}^k,0) \subset (\mathbb{C}^{k'},0)$). We claim that $\rho_f(x) : T_0(S) \to T_{X_0,x_0}$ is injective: if $v \in T_0(S)$ is mapped to zero, then it follows from the definitions that v is in the image of $\partial f(x)$ and hence $v = 0$.

Write τ for $\tau(X_0,x_0)$ and choose $\zeta_{k+1},\ldots,\zeta_\tau \in \theta(f)$ such that $\{\zeta_{k+1},\ldots,\zeta_\tau\}$ projects onto a $\mathbb{C}$-basis of Coker $\rho_f(x)$. Use the coefficients $\zeta_{\lambda\kappa}$ of ζ_κ $(\zeta_\lambda = \Sigma_{\kappa=1}^k \zeta_{\lambda\kappa} \frac{\partial}{\partial t_\kappa})$ to define a map $F : (X\times\mathbb{C}^{\tau-k}, x\times 0) \to (\mathbb{C}^k\times\mathbb{C}^{\tau-k}, 0\times 0)$, $F(z,u) = (g(z,u),u)$ where $g_\kappa(z,u) = f_\kappa(z) + \Sigma_{\lambda=\kappa+1}^{\tau} \zeta_{\lambda\kappa}u_\lambda$. It is clear that F defines a deformation of (X_0,x_0). We assert that $\rho_F(x\times 0)$ is an isomorphism. For reasons of dimension, it is enough to show that $\rho_F(x\times 0)$ is surjective. The equality

$$\partial F\left(\frac{\partial}{\partial u_\lambda}\right) = \Sigma_{\kappa=1}^k \zeta_{\lambda\kappa} \frac{\partial}{\partial t_\kappa} + \frac{\partial}{\partial u_\lambda}$$

(where on the left, $\frac{\partial}{\partial u_\lambda}$ is viewed as an element of $\theta_{X\times\mathbf{C}^{\tau-k},x\times 0}$, while on the right it belongs to $\theta_{S\times\mathbf{C}^{\tau-k},\theta\times 0}$) shows that the image of ζ_λ in T_F is just $-\rho_F(\frac{\partial}{\partial u_\lambda})$. On the other hand, $\rho_F(\frac{\partial}{\partial t_\kappa}) = \rho_f(\frac{\partial}{\partial t_\kappa})$ by (6.2). So $\rho_F(x\times 0)$ is surjective.

To prove the last part, just observe that had we started with an arbitrary deformation (ι,f) of (X_0,x_0), the above construction would have led to a deformation (ι,F) with ρ_F surjective and an embedding $(\iota,f) \subset (\iota,F)$.

Proof of (6.5). (i) Let (ι,f) be a versal deformation. Following (6.6) there exists a deformation (ι',f') of (X_0,x_0) with $\rho_{f'}$ surjective. By the defining property of versality there exists a morphism (g',g) from (ι',f') to (ι,f). Then $\rho_f(x)\circ\partial g(x') = \rho_{f'}(x')$ by (6.2) and hence $\rho_f(x)$ is surjective.

Next let (ι,f) be a deformation with ρ_f surjective. We want to show that there is a morphism from any deformation (ι',f') to (ι,f). By lemma (6.6), (ι',f') maps to a deformation (ι'',f'') with $\rho_{f''}$ surjective. By multiplying (ι'',f'') with a trivial factor (replace f'' by $f''\times\mathbf{1}_{\mathbf{C}^N}$ for some N), we may assume that $\ell := \dim(S'',s'') - \dim(S,s) \geq 0$. Now $f\times\mathbf{1}_{\mathbf{C}^\ell}$ also has a surjective Kodaira-Spencer map and its base dimension is that of f''. It now follows from (1.7) and prop. (6.3) that (ι'',f'') and $(\iota,f\times\mathbf{1}_{\mathbf{C}^\ell})$ are isomorphic deformations. There is also a natural projection of $(\iota,f\times\mathbf{1}_{\mathbf{C}^\ell})$ onto (ι,f). Composing these morphisms gives a morphism from (ι',f') to (ι,f).

(ii) Immediate from (i), (1.7) and prop. (6.3).

(iii) and (iv) Let (ι,f) be a deformation such that $\rho_f(x)$ is an isomorphism. If (ι',f') is any deformation, then there exists by (i) a morphism (g,g') from (ι',f') to (ι,f). Since $\rho_f(x)\circ\partial g(x') = \rho_{f'}(x')$ and

$\rho_f(x)$ is invertible, it follows that $\partial g(x')$ is unique. So (ι,f) is miniversal.

Conversely, suppose (ι,f) miniversal. Then $\rho_f(x)$ is surjective by (i). Let (ι',f') be a deformation for which $\rho_{f'}(x')$ is an isomorphism (exists by lemma (6.6)). Then (ii) implies that (ι,f) is isomorphic to $(\iota',f'\times 1_{\mathbf{C}^\ell})$ for some ℓ. We must show that $\ell = 0$. The maps $(S'\times C^\ell, s'\times 0)\circlearrowleft$ given by the identity resp. the projection onto $S'\times\{0\}$ both induce the deformation $f'\times 1_{\mathbf{C}^\ell,0}$. Because of miniversality, the differentials of these maps must agree, which can only be the case if $\ell = 0$.

(6.7) In the proof of (6.6) we gave a construction of a miniversal deformation. Let us redo this construction in terms of coordinates, just to show how easily this is accomplished in practice. Let be given a germ $f : (\mathbf{C}^{n+k},0) \to (\mathbf{C}^k,0)$ defining an icis of dim n such that $df(0) = 0$. Then $T_{f,0}$ is the cokernel of the $\mathcal{O}_{\mathbf{C}^{n+k},0}$ homomorphism

$$\partial f = \left(\frac{\partial f_\kappa}{\partial z_\nu}\right) : \mathcal{O}^{n+k}_{\mathbf{C}^{n+k},0} \to \mathcal{O}^k_{\mathbf{C}^{n+k},0}$$

and $T_{(X_0,0)}$ is its reduction mod $(f_1,\ldots,f_k)\mathcal{O}_{\mathbf{C}^{n+k},0}$. Choose column vectors $\phi_1,\ldots,\phi_{\tau-k} \in (m_{\mathbf{C}^{n+k},0})^k$ such that $\phi_1,\ldots,\phi_{\tau-k}$ map onto a $\mathbf{C}$-basis of

$$(m_{\mathbf{C}^{n+k},0})^k/[\partial f(\mathcal{O}^{n+k}_{\mathbf{C}^{n+k},0}) + (f_1,\ldots,f_k)\mathcal{O}^k_{\mathbf{C}^{n+k},0}].$$

(An efficient way to find such $\phi_1,\ldots,\phi_{\tau-k}$ is first to determine an $\ell \in \mathbf{N}$ such that

$$m^\ell_{\mathbf{C}^{n+k},0} \subset C_{f,0} + (f_1,\ldots,f_k)\mathcal{O}_{\mathbf{C}^{n+k},0} + m^{\ell+1}_{\mathbf{C}^{n+k},0}. \tag{6.7.a}$$

Nakayama's lemma implies that we may then omit the term $m^{\ell+1}_{\mathbf{C}^{n+k},0}$. Since

$(C_{f,0})^k \subset \partial f(\mathcal{O}_{\mathbb{C}^{n+k},0}^{n+k})$, this allows us to calculate modulo $m^{\ell}_{\mathbb{C}^{n+k},0}$. Mather (1968) gives a function $\ell(n,k;d)$ with the property that if $f_1,\ldots,f_k$ are polynomials of degree $\leq d$ (6.7a) holds with $\ell = \ell(n,k;d)$ if and only if $f = (f_1,\ldots,f_k)$ defines an icis.) The germ $(\mathbb{C}^{n+k}\times\mathbb{C}^{\tau-k},0\times 0) \to (\mathbb{C}^k\times\mathbb{C}^{\tau-k},0\times 0)$, $(z,u) \to (f(z)+u_1\phi_1(z)+\ldots+u_{\tau-k}\phi_{\tau-k}(z),u_1,\ldots,u_{\tau-k})$ is then miniversal.

Example 2. A miniversal deformation of the icis given by $f : (\mathbb{C}^{n+2},0) \to (\mathbb{C}^2,0)$, $z \to (z_1^2+\ldots+z_{n+2}^2,\lambda_1 z_1^2+\ldots+\lambda_{n+2}z_{n+2}^2)$, where $\lambda_1,\ldots,\lambda_{n+2}$ are all distinct (see example 1 of Ch. 5) is $F : (\mathbb{C}^{n+2}\times\mathbb{C}^{n-1}\times\mathbb{C}^{n+2},0) \to (\mathbb{C}\times\mathbb{C}\times\mathbb{C}^{n-1}\times\mathbb{C}^{n+2},0)$, $F(z,u,v) = (f_1(z),g(z,u,v),u,v)$, where $g(z,u,v) = f_2(z) + u_1 z_1^2+\ldots+u_{n-1}z_{n-1}^2 + v_1 z_1+\ldots+v_{n+2}z_{n+2}$. Notice that in this case $\tau = \mu = 2n+3$.

(6.8) If $(\tilde{g},g)$ is a morphism from a deformation (ι',f') to a deformation (ι,f), then one easily verifies that the pull-back of a good representative $f : X \to S$ of f over any representative $g : S' \to S$ of g with S' contractible is a good representative of f'. This implies that a Milnor fibre of the latter is diffeomorphic to a Milnor fibre of f. In view of the existence of versal deformations, it follows that any two Milnor fibres associated to an icis are diffeomorphic (under a diffeomorphism whose restriction to the boundary is in a preferred isotopy class). So the Milnor number is indeed an invariant of a given icis.

(6.9) Our definition of a deformation is a much narrower one than is customary. Normally one defines a deformation of an isolated singularity (X_0,x_0) as a pair (ι,f) where now $f : (X,x) \to (S,s)$ is a flat morphism of analytic space germs (and ι still an isomorphism of (X_0,x_0) onto the closed fibre (X_s,x)). The condition "f is flat" means that f* makes of

$\mathcal{O}_{X,x}$ a flat $\mathcal{O}_{S,s}$-module. The main geometric consequence of this condition is that it implies the existence of good (proper) representatives $f : X \to S$ of f: outside the critical locus C of f (which is finite over S), f is locally a projection with nonsingular fibres (but S may be singular). For such representatives, we may have smooth fibres (in which case f is called a *smoothing* of (X_0,x_0)) but not necessarily. With this more general notion of a deformation, a Kodaira-Spencer map can still be defined such that an analogue of (6.3) holds. In particular, miniversal deformations exist and are unique. There exist isolated singularities which cannot be deformed at all: they are so to speak their own miniversal deformations. Such singularities are called rigid. For instance the quotient singularities of dim ≥ 3 are rigid (Schlessinger, 1971). (It is conjectured that there are no rigid singularities in dim 2. Pinkham (1974) showed that for example 4 of ch. I with d = 4 the base of a miniversal deformation has two irreducible components of dim 3 and 1. For both components the general fibre is nonsingular, but they are not homotopy equivalent: one is simply connected while the other has fundamental group of order two. So neither the existence, nor the essential uniqueness of the Milnor fibre is guaranteed anymore.

For our purposes we can do with a somewhat less general definition of a deformation. *We shall only allow map germs* $f : (X,x) \to (S,s)$ *as deformations of an n-dim icis for which* (S,s) *is nonsingular and* (X,x) *is a complete intersection of dimension* $n + \dim(S,s)$. Then (X,x) is pure-dimensional, so that good representatives as constructed in (2.7) exist. The following proposition shows that such map germs are flat and that for this extension of our deformation category the notion of a versal deformation is unaffected.

(6.10) *Proposition.* Let (X,x) be a germ of an analytic space in $\mathbb{C}^N$ and let $f : (X,x) \to (\mathbb{C}^k,0)$ be a morphism such that (X_0,x) is a complete intersection of dim n. Then f is flat if and only if (X,x) is a complete intersection of dim n+k. In either case, f fits in a cartesian diagram (= a morphism of deformations)

$$\begin{array}{ccc} (X,x) & \subset & (\mathbb{C}^N,0) \\ f \downarrow & & \downarrow F \\ \mathbb{C}^k & \subset & \mathbb{C}^k \times \mathbb{C}^{N-n-k} \end{array}$$

Proof. Suppose f flat and assume first that $k = 1$. The flatness then just means that f is not a zero divisor in $\mathcal{O}_{X,x}$. Notice that (X_0,x) is defined as a subgerm of (X,x) by the principal ideal $f\mathcal{O}_{X,x}$. Let $\tilde{f} \in m_{\mathbb{C}^N,x}$ project onto f. Since (X_0,x) is a complete intersection of dim n, $I_{X_0,x}$ is generated by a regular sequence $F_1,\ldots,F_{N-n}$ in $\mathcal{O}_{\mathbb{C}^N,x}$. Clearly, $\tilde{f} \in I_{X_0,x}$. We cannot have $\tilde{f} \in m_{\mathbb{C}^N,x} I_{X_0,x}$, for this would mean that $I_{X_0,x}/I_{X,x} = m_{\mathbb{C}^N,x} I_{X_0,x}/I_{X,x}$. Nakayama's lemma then implies $I_{X_0,x} = I_{X,x}$ and so $f = 0$ in $\mathcal{O}_{X,x}$, which of course contradicts the flatness of f. Hence $\tilde{f} = \phi_1 F_1 + \ldots + \phi_{N-n} F_{N-n}$ for certain $\phi_\nu \in \mathcal{O}_{\mathbb{C}^N,x}$, not all in $m_{\mathbb{C}^N,x}$. If for instance $\phi_1 \notin m_{\mathbb{C}^N,x}$, then we clearly may replace F_1 by $\tilde{f}$. Modify $F_2,\ldots,F_{N-n}$ by adding multiples of $\tilde{f}$ to make them all belong to $I_{X,x}$. Thus $F_2,\ldots,F_{N-n}$ is now a regular sequence in $I_{X,x}$. We prove that $F_2,\ldots,F_{N-n}$ generate $I_{X,x}$. If $G \in I_{X,x}$ then $G = \psi_1 F_1 + \ldots + \psi_{N-n} F_{N-n}$ for certain ψ_ν. If we reduce mod $I_{X,x}$ we find that $G \equiv \psi_1 \tilde{f} \bmod I_{X,x}$. Since f is not a zero divisor in $\mathcal{O}_{X,x}$, it follows that $\psi_1 \in I_{X,x}$. So $I_{X,x} = \tilde{f} I_{X,x} + (F_2,\ldots,F_{N-n})\mathcal{O}_{\mathbb{C}^N,x}$ and another application of Nakayama's lemma proves that $I_{X,x} = (F_2,\ldots,F_{N-n})\mathcal{O}_{\mathbb{C}^N,x}$. So (X,x) is a complete intersection of dim n+1. The general case $(k \geq 1)$ proceeds with induction on k: if $f : (X,x) \to (\mathbb{C}^k,0)$ is flat, then so is $(f_k^{-1}(0),x) \to (\mathbb{C}^{k-1},0)$.

The induction hypothesis tells us that $(f_k^{-1}(0),x)$ is a complete intersection of dim n+k-1 and by the preceding result, (X,x) is then a complete intersection of dim n+k.

Now suppose that (X,x) is an $(n+k)$-dimensional complete intersection. Then there exists a regular sequence $F_{k+1},\dots,F_{N-n}$ in $\mathcal{O}_{\mathbb{C}^N,x}$ defining (X,x). Let $F_\kappa \in \mathcal{O}_{\mathbb{C}^N,x}$ project onto f_κ $(\kappa = 1,\dots,k)$. Then $F := (F_1,\dots,F_{N-n}) : (\mathbb{C}^N,x) \to (\mathbb{C}^{N-n},0)$ and f clearly form a cartesian diagram. The flatness of both f and F now follows (Matsumura (1980) (20.D)). (Although the result stated there pertains to polynomial rings, the same proof works.)

6.D *Some analytic properties of versal deformations*

We first show that a versal deformation is Thom-transversal, so that it has all the nice properties described in (4.7) and (4.11).

(6.11) *Proposition.* Let $f : (X,x) \to (S,0)$ be a versal deformation of an icis. Then f is Thom-transversal.

Proof. By (4.12) we can extend f to a Thom-transversal map $F : (X\times\mathbb{C}^\ell, x\times 0) \to (S\times\mathbb{C}^\ell, 0\times 0)$ of the form $F(z,u) = (g(z,u),u)$, with $g(z,0) = f(z)$. Then $\rho_F(x)$ is surjective because its restriction $\rho_f(x)$ is. So F is versal and hence by (6.5) analytically equivalent to $f\times\mathbf{1} : (X\times\mathbb{C}^\ell, x\times 0) \to (S\times\mathbb{C}^\ell, 0\times 0)$. It follows that $f\times\mathbf{1}$ is Thom-transversal. This implies that f is Thom-transversal.

Next we derive a Cohen-Macaulay property of T_f. This has a remarkable consequence for the discriminant of a versal deformation.

(6.12) *Proposition*. Let $f : (\mathbb{C}^{n+k},x) = (X,x) \to (S,0) = (\mathbb{C}^k,0)$ define an icis of dim n. Then $T_{f,x}$ is a Cohen-Macaulay module of dim k-1, both viewed as an $\mathcal{O}_{X,x}$-module and as an $\mathcal{O}_{S,0}$-module.

Proof. We have a presentation of $T_{f,x}$ by free $\mathcal{O}_{X,x}$-modules:

$$\theta_{X,x} \xrightarrow{\partial f} \theta(f) \to T_{f,x} \to 0$$

of rank n+k and k respectively. By definition, $C_{f,x}$ is the ideal generated by the determinants of the k×k-minors of ∂f. Since $C_{f,x}$ defines a locus of dim k-1 (the critical locus), it follows from Buchsbaum & Rim (1964), Cor. 2.7, that $T_{f,x}$ is Cohen-Macaulay as an $\mathcal{O}_{X,x}$-module. Since $C_{f,x}$ annihilates $T_{f,x}$, $T_{f,x}$ is then also Cohen-Macaulay when viewed as an $\mathcal{O}_{C,x} = \mathcal{O}_{X,x}/C_{f,x}$-module. Since $\mathcal{O}_{C,x}$ is a finite $\mathcal{O}_{S,0}$-module, $T_{f,x}$ is also Cohen-Macaulay as an $\mathcal{O}_{S,0}$-module (Serre (1975) Ch. IV, prop. 11).

(6.13) *Corollary*. Suppose $f : (X,x) \to (S,0)$ versal. Then we have an exact sequence of $\mathcal{O}_{S,0}$-modules

$$0 \to \theta_{S,0}\langle D\rangle \to \theta_{S,0} \xrightarrow{\rho_f} T_{f,x} \to 0$$

and $\theta_{S,0}\langle D\rangle$ is a *free* $\mathcal{O}_{S,0}$-module of $\theta_{S,0}$ of rank k.

Here $\theta_{S,0}\langle D\rangle$ denotes the set of $\eta \in \theta_{S,0}$ which preserve the ideal $I_{D,0}$. So this sequence identifies $T_{f,x}$ with a kind of normal bundle to the discriminant. The freeness of $\theta_{S,0}\langle D\rangle$ in the hypersurface case was first proved by Saito by a different method. We will use Cor. (6.13) in ch. 9 when we discuss the period mapping.

Proof. Since $T_{f,x}$ is a Cohen-Macaulay $\mathcal{O}_{S,0}$-module of dim k-1, its homological dimension (as an $\mathcal{O}_{S,0}$-module) is 1 (Serre (1975) Ch. IV, prop. 21). Since $\theta_{S,0}$ is a free $\mathcal{O}_{S,0}$-module and $\rho_{f,x}$ is surjective, it follows that $\mathrm{Ker}(\rho_{f,x})$ is a free submodule of $\theta_{S,0}$. So the corollary will follow from

(6.14) *Lemma*. Let $f : (X,x) \to (S,0)$ be Thom-transversal. Then $\mathrm{Ker}(\rho_{f,x}) = \theta_{S,0}\langle D\rangle$. In other words, $\eta \in \theta_{S,0}$ lifts over f if and only if η preserves the discriminant of f.

Proof. Let $\eta \in \mathrm{Ker}(\rho_{f,x})$. This means that there exists a $\xi \in \theta_{X,x}$ with $\partial f(\xi) = f^*(\eta)$. If we evaluate this equation in a critical point of f which maps to a regular point of the discriminant and use the normal form (4.2.iv) then we find that η is tangent to the discriminant at its regular points. So if δ is a generator of $I_{D,0}$, then $\eta(\delta)$ vanishes on $(D,0)$. As $(D,0)$ is reduced, it follows that $\eta(\delta) \in I_{D,0}$.

Suppose now $\eta \in \theta_{S,0}\langle D\rangle$. By (4.7), $f : (C,x) \to (D,0)$ is a normalization, and so $\eta|D$ lifts to a derivation ξ_C of (C,x). (This is geometrically seen as follows: η generates a one-parameter family of isomorphisms: $(D\times\mathbb{C},0\times 0) \to (D,0)$. By the universal property of normalization this lifts to a one-parameter family of isomorphisms $(C\times\mathbb{C},x\times 0) \to (C,x)$. The derivative of this lift gives the lift of η.) We may extend ξ_C to a $\xi \in \theta_{X,x}$. Clearly ξ preserves (C,x). Now consider $\zeta := -\partial f(\xi)+f^*(\eta) \in \theta(f)_x$. By construction, $\zeta \in C_{f,x}\theta(f)_x$. Cramer's rule implies that $C_{f,x}\theta(f)_x$ is contained in $\partial f(\theta_{X,x})$. So there exists a $\xi' \in \theta_{X,x}$ with $\zeta = \partial f(\xi')$. Hence $f^*(\eta) = \partial f(\xi+\xi')$, which proves that $\eta \in \mathrm{Ker}(\rho_{f,x})$.

(6.15) Let $f : X \to S$ be a good representative of a germ $(\mathbb{C}^{n+k},x) \to (\mathbb{C}^k,0)$ which defines an icis. Then we can sheafify $\theta(f)_x$ and $T_{f,x}$ by putting

$$\theta(f) := \mathcal{O}_X \otimes_{\mathcal{O}_S} \theta_S$$

$$T_f := \mathrm{Coker}(\partial f : \theta_X \to \theta(f)).$$

As in the punctual case, T_f is in fact a (coherent) $\mathcal{O}_C$-module. Since $f : C \to S$ is finite, f_*T_f is a coherent $\mathcal{O}_S$-(in fact $\mathcal{O}_D$-)module. Notice that the fibre $f_*T_f/m_{S,s}f_*T_f$ of f_*T_f in $s \in S$ may be identified with the direct sum of the $T_{X_s,y}$, $y \in X_{s,sing}$. By taking S sufficiently small, we may suppose that T_f is generated by $\tau(X_0,x) = \dim_{\mathbb{C}} T_{X_0,x}$ elements. So we then have

$$(6.15.1) \qquad \Sigma_{y \in X_{s,sing}} \tau(X_s,y) \le \tau(X_0,x).$$

In this context, the Kodaira-Spencer map is an $\mathcal{O}_S$-homomorphism

$$\rho_f : \theta_S \to f_*T_f$$

Now suppose that f is versal in x. Then $\rho_{f,x}$ is surjective and by shrinking S, we may suppose that ρ_f is surjective. This implies that f is versal in any point $y \in X$, but we get more than that: for any $s \in S$, we find a surjection

$$\oplus_{y \in X_{s,sing}} \rho_f(y) : T_s(S) \to \oplus_{y \in X_{s,sing}} T_{X_s,y}$$

This fact can be interpreted geometrically as follows. If $y \in C$, let $M_y \subset X$ denote the set of $y' \in X$ where f defines a singularity isomorphic to (X_s,y), where $s = f(y)$. It can be shown that M_y is an analytic Zariski-constructible submanifold of X such that $f|M_y$ is an immersion. Moreover, the image of $T_y(M_y)$ under ∂f is just the kernel of the map $\rho_f(y) : T_s(S) \to T_{X_s,y}$. In particular, f is miniversal in y if and only if y is an isolated point of M_y. Now the surjectivity of $\oplus\rho_f(y) : T_s(S) \to \oplus T_{X_s,y}$ just says that the images of the germs (M_y,y), $y \in X_{s,sing}$ under f are in general position ('multitransversal') in (S,s). See for instance Mather (1969, 1970) and Teissier (1972).

7 VANISHING LATTICES, MONODROMY GROUPS AND ADJACENCY

Section B of this chapter is concerned with a major discrete invariant of an icis, namely its vanishing lattice. This consists of the data: H_n(Milnor fibre), its intersection pairing and the set of vanishing cycles in H_n(Milnor fibre). In order to derive some of its properties we need to know some simple facts concerning fundamental groups of hypersurface complements. These are proved in §A. Section C deals with the relative complexity of a singularity as formalized by the notion of adjacency. The adjacency relation among two singularities implies an inclusion relation among their vanishing lattices, so that the vanishing lattice is usually a good absolute measure for the complexity of a singularity. This is illustrated by the partial classification of icis's described in §D.

7.A *The fundamental group of a hypersurface complement*

Let $T \subset \mathbf{C}^{k-1}$ be open and contractible, $\Delta \subset \mathbf{C}$ a disc of finite radius η and put $S = T\times\bar{\Delta}$. Suppose that we are given an analytic hypersurface $D \subset T\times\Delta$, which is closed in $T\times\bar{\Delta}$. Then the restriction of the projection $\pi : S \to T$ to D is finite. Let $B \dot{\subset} T$ denote the discriminant of $\pi|D$ (the set of $t \in T$ for which the number of points of D_t is not maximal).

This is a proper subvariety of T. Choose $t \in T-B$ and put $s = (t,\eta)$.

(7.1) *Lemma*. The inclusion $i : S_t-D_t \subset S-D$ induces a surjection of fundamental groups.

Proof. Over T-B, $\pi : S-D \to T$ is a locally trivial fibre bundle which admits the section $\sigma : t \in T-B \to (t,\eta) \in S-D$. This section splits the associated exact homotopy sequence (e.g. Whitehead (1978)), so that we have an isomorphism

$$(i_*,\sigma_*) : \pi(S_t-D_t,s)\times\pi(T-B,t) \to \pi((S-D)_{T-B},s).$$

Now let ω be a loop in S-D based at s. Since B is of real codim ≥ 2 in T, a small homotopy of ω will make $\pi\circ\omega$ disjoint from B. By what we just proved, ω is then homotopic in $(S-D)_{T-B}$ to a loop of the form $\omega_1*\omega_2$, where ω_1 is a loop in S_t-D_t and ω_2 a loop in $T\times\{\eta\}$. Since $T\times\{\eta\}$ is simply connected, it follows that in S-D, ω is homotopic to ω_1.

Denote the distinct points of D_t by $s_1,\ldots,s_m$ (so that m is the degree of $\pi|D$). Around each s_μ we choose a simple loop ω_μ in S_t-D_t based at s around s_μ. More precisely, ω_μ is the composite of a piecewise linear path γ_μ in S_t-D_t, a simply positively oriented loop ℓ_μ around s_μ

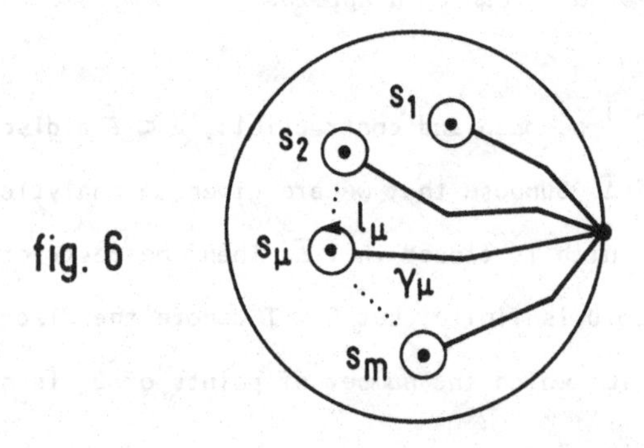

fig. 6

and γ_μ^{-1}. We do this in such a way that each ω_μ is without self-intersection and that distinct loops only intersect in s. The homotopy classes of $\omega_1,\ldots,\omega_m$ freely generate $\pi(S_t-D_t,s)$. Then by (7.1) they also generate $\pi(S-D,s)$.

This lemma is a special and simple case of a Zariski type theorem due to Hamm & Lê which asserts that under appropriate genericity assumptions (concerning the shape of S and the choice of an affine subspace $P \subset \mathbb{C}^k$) the complement S-D is obtained up to homotopy from its intersection with P by attaching finitely many cells of dim > dim P to $(S-D) \cap P$. So $(S-D) \cap P \subset (S-D)$ then induces a surjection resp. bijection on fundamental groups if P is a line resp. a plane.

(7.2) *Lemma*. Assume D irreducible. Then the homotopy classes of $\omega_1,\ldots,\omega_m$ in $\pi(S_t-D_t,s)$ are in a single conjugacy class.

Proof. Since D is irreducible, D_{reg} is connected (e.g. Narasimhan (1966)) and so s_1 and s_μ can be connected by an arc γ in D_{reg}. We can push this arc away from D so that the resulting arc $\tilde{\gamma}$ connects the end point of γ_1

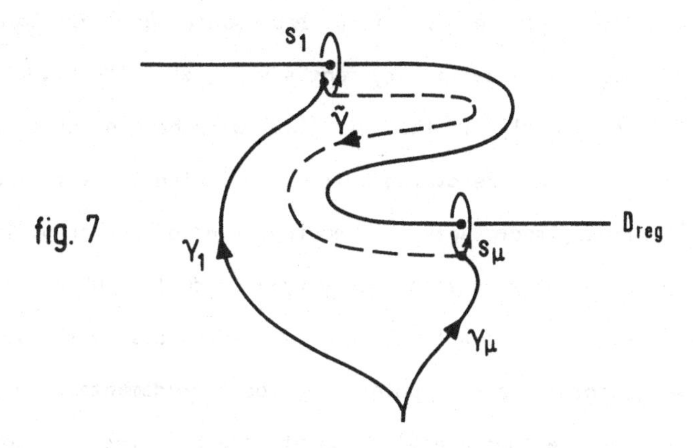

fig. 7

with the end point of γ_μ. It is easy to see that ω_1 and $\gamma_1 * \tilde{\gamma} * \ell_\mu * \tilde{\gamma}^{-1} * \gamma_1^{-1}$ (with some order of the parentheses) are homotopic. The homotopy class of the latter is just $[\omega_1]$ conjugated with $[\gamma_1 * \tilde{\gamma} * \gamma_\mu^{-1}]$.

(7.3) We will be concerned with hypersurface germs $(D,0) \subset (\mathbb{C}^k,0)$, rather than with actual hypersurfaces. To see that the above results retain their meaning in this context, we need the notion of local fundamental group.

Assume we are given a subspace D of a topological space S and that p is a point of D. Suppose that p possesses a neighbourhood basis $\{U_i\}_{i \in I}$ in S with the property that $U_{i_1} \subset U_{i_2}$ implies that $U_{i_1} \cap (S-D) \subset U_{i_2} \cap (S-D)$ is a homotopy equivalence. Then we claim that the homotopy type of $U_i \cap (S-D)$ is not only independent of $i \in I$, but also of the neighbourhood basis itself. To see this, let $\{V_j\}_{j \in J}$ be a neighbourhood basis with the same property and pick $i_1, i_2 \in I$ and $j_1, j_2 \in J$ such that $V_{j_2} \subset U_{i_2} \subset V_{j_1} \subset U_{i_1}$. If we intersect this chain of inclusions with S-D, then the composition of any two consecutive inclusions is a homotopy equivalence. This implies that each of the inclusions is a homotopy equivalence. If U is a member of such a neighbourhood basis, we say that $U \cap (S-D)$ *represents the local homotopy type of* S-D *at* p. By the preceding argument any two such are naturally homotopy equivalent. Clearly, if the pair (S,D) with base point p happens to be (homeomorphic to) a cone over a pair (S_1,D_1) with its vertex as base point, then S-D already represents the homotopy type of S-D at p. In particular, the notion of a local homotopy type is defined whenever the pair of spaces involved is locally conical (as is the case with analytic sets). Since we cannot fix a base point, the local fundamental group of S-D at p (defined as the fundamental group of some $U \cap (Y-Z)$ as above) is only defined "up to inner automorphism".

The case which concerns us here is when D is an analytic hypersurface in a neighbourhood Ω of $0 \in \mathbb{C}^k$ such that D meets the last coordinate axis in {0} only. Using a Whitney stratification of D it can be shown that there exists a neighbourhood S satisfying the hypotheses of (7.1) such that S-D represents the homotopy type of S-D at 0.

7.B *The monodromy group*

(7.4) We apply the results of the preceding section to the case where D is a discriminant. We start off with a versal deformation $f : (\mathbb{C}^{n+k},x) \to (\mathbb{C}^k,0)$ of an icis (X_0,x). Recall that f has then a reduced and irreducible discriminant (D,0). We suppose that the last coordinate axis is not in the tangent cone of (D,0) and we choose a good proper representative $f : \overline{X} \to S = T\times\overline{\Delta}$ as in (5.4) such that S-D represents the homotopy type of S-D at 0. In view of the essential uniqueness of a versal deformation (up to a trivial factor), the homotopy type of S-D at 0 only depends on (X_0,x). We let $s = (t,\eta)$, $\omega_1,\ldots,\omega_m$ etc. be as in the previous section (so that m is the multiplicity $\nu(D,0)$ of D in 0), but we now make the additional assumption that the $\gamma_1,\ldots,\gamma_m$ have been chosen in such a way that the direction $\dot{\gamma}_\mu(0)$ increases with μ.

In (2.C) we described the geometric monodromy representation; this was a homomorphism from $\pi(S-D,s)$ to the group of relative isotopy classes of $(\overline{X}_s,\partial\overline{X}_s)$. Passing to homology gives a representation of $\pi(S-D,s)$ on the exact homology sequence of the pair $(\overline{X}_s,\partial\overline{X}_s)$. Since $\overline{X}_s$ has the homotopy type of a bouquet of n-spheres, $\overline{X}_s$ has only interesting homology in dim n. Likewise, $(\overline{X}_s,\partial\overline{X}_s)$ has only interesting homology in dim n because by Lefschetz duality $H_i(\overline{X}_s,\partial\overline{X}_s) = H^{2n-i}(\overline{X}_s)$, which is 0 if $i \neq n,2n$ and is the **Z**-dual $H_n(\overline{X}_s)^*$ of $H_n(\overline{X}_s)$ if $i = n$ (by the universal coefficient theorem). So there remains an exact sequence

$$0 \to H_n(\partial\overline{X}_s) \to H_n(\overline{X}_s) \overset{j_*}{\to} H_n(\overline{X}_s)^* \to \widetilde{H}_{n-1}(\partial\overline{X}_s) \to 0$$

Clearly $\pi(S-D,s)$ acts trivially on the outer terms and the action on $H_n(\overline{X}_s)^*$ is the contragredient of the one on $H_n(\overline{X}_s)$: $\sigma.\phi = \phi\circ\sigma^{-1}$. So the homological monodromy is already determined by the representation on $H_n(\overline{X}_s)$. The image of this representation in the automorphism group of $H_n(\overline{X}_s)$ is called the *monodromy group* (of f) and we shall denote it by Γ. Notice that we have an intersection form on $H_n(\overline{X}_s)$ defined by $(x.y) = j_*(x)(y)$. This form is symmetric or skew according to whether n is even or odd. Since Γ acts trivially on $H^{2n}(\overline{X}_s,\partial\overline{X}_s)$, (.) is left invariant by Γ.

Each path γ_μ determines a vanishing cycle $\delta_\mu \in H_n(\overline{X}_s)$ (up to sign) and following the Picard-Lefschetz formula, the image of $[\omega_\mu]$ in Γ is the pseudo-reflection given by

$$\sigma_\mu(v) = v - (-1)^{\frac{1}{2}n(n-1)}(v.\delta_\mu)\delta_\mu$$

(7.5) *Proposition*. The vanishing cycles $\delta_1,\ldots,\delta_m$ generate $H_n(\overline{X}_s)$ (freely if (X_0,x) is a hypersurface germ) and the corresponding pseudo-reflections $\sigma_1,\ldots,\sigma_m$ generate Γ and are in a single conjugacy class of Γ. We cannot decompose $\{1,\ldots,m\}$ into disjoint nonvoid subsets I,J such that $(\delta_i.\delta_j) = 0$ if $i \in I$ and $j \in J$.

Proof. The first assertion follows from (5.11), (7.1) and (7.2). If $\{I,J\}$ were a decomposition as in the proposition, then the σ_i, $i \in I$ would commute with the σ_j, $j \in J$, so that Γ decomposes accordingly as $\Gamma_I \times \Gamma_J$.

Clearly, no σ_i, $i \in I$ is then conjugate to a σ_j, $j \in J$.

We shall refer to an ordered set $(\delta_1,\ldots,\delta_m)$ thus obtained as a *distinguished system of generators for* $H_n(\overline{X}_s)$. In the hypersurface case this will also be called a *distinguished basis for* $H_n(\overline{X}_s)$.

(7.6) *Corollary*. If D has multiplicity m at $s \in S$, then (D,s) is an irreducible germ and X_s has precisely one singularity.

Proof. Put $t = \pi(s)$. Since $\pi : D \to T$ is of degree m, s is the unique pre-image of t in D. Consider the finite map germs

$$(C,C_s) \xrightarrow{f} (D,s) \xrightarrow{\pi} (T,t_0)$$

of which the first is a normalization and the second is of degree m. If C_s is not a singleton, let $\{C_1,C_2\}$ be a nontrivial partition of C_s. Choose $s' = (t',\eta')$ close to s such that $\{t'\}\times\Delta$ meets D in m distinct (regular) points. The above procedure gives m vanishing cycles $\delta_1,\ldots,\delta_m$. The set $\{1,\ldots,m\}$ decomposes according to the partition $\{C_1,C_2\}$ into $\{I_1,I_2\}$ ($\mu \in I_k$ if and only if δ_μ "vanishes" at a point of C close to C_k) and it is geometrically clear that for $i \in I_1$, $j \in I_2$, δ_i and δ_j will not intersect. This contradicts the previous proposition.

(7.7) The set $D^{(m)}$ of $s \in D$ where D has multiplicity m is a subvariety of D. In the hypersurface case, this is by (5.11) and (7.6) also the set of $s \in D$ over which there is precisely one singular point with Milnor number $m = \mu(X_0,x)$. Teissier (1972) has shown that in this case $f : C_{D^{(m)}} \to D^{(m)}$ is an isomorphism.

If γ is a path in S from s to a regular point of D such that $\gamma|[0,1)$ stays in S-D, then the construction of Ch. 3 gives us a vanishing cycle $\pm \delta_\gamma$ (the sign is not defined) in $H_n(\overline{X}_s)$. We shall denote the collection of all such classes by $\Delta \subset H_n(\overline{X}_s)$. It is not known whether there is always a basis of $H_n(\overline{X}_s)$ consisting of vanishing cycles.

(7.8) *Proposition*. The set Δ of vanishing cycles forms a single Γ-orbit, except when (X_0,x) is a quadratic singularity of odd dimension (in this case Γ is trivial and there are two vanishing cycles $\pm \delta$).

Proof. Let $\delta,\delta' \in \Delta$. The sort of argument used in the proof of lemma (7.2) (based on the connectedness of D_{reg}) shows that there is $\sigma \in \Gamma$ which sends δ' to $\pm \delta$. If n is even, then $\sigma_\delta(\delta) = -\delta$, so in this case it is clear that $\delta' \in \Gamma.\{\delta\}$. If n is odd and (X_0,x) is not an ordinary double point, then there exists a $\delta'' \in \Delta$ with $(\delta.\delta'') = \pm 1$ (we shall prove this later (in (7.18)) using the fact that X_0 deforms into an A_2-singularity). A straightforward computation shows that then $\sigma_{\delta''}\sigma_\delta\sigma_\delta\sigma_{\delta''}$ maps δ to $-\delta$. So in this case, $\delta' \in \Gamma.\{\delta\}$ also.

(7.9) In case $n = 0$, Γ can be identified with a permutation group of the finite set X_s. We claim that it is the full permutation group, for

(i) Γ is generated by transpositions $\sigma_1,\ldots,\sigma_m$ which are all conjugate in Γ.

(ii) for each $y \in X_s$ there is a μ with $\sigma_\mu(y) \neq y$.

It is an exercise in group theory to show that these two properties imply that $\Gamma = \mathrm{Aut}(X_s)$.

Let us now consider the more interesting case when $n > 0$. It will be convenient to have some notation first. We refer to $H_n(\overline{X}_s)$ endowed with the intersection form (.) as a *Milnor lattice* of (X_0,x) and we denote it V. If we include Δ as a datum, then we shall call it a *vanishing lattice*. We write $j : V \to V^*$ for the adjoint of the intersection form, so that $\mathrm{Ker}(j) \cong H_n(\partial\overline{X}_s)$ and $\mathrm{Coker}(j) \cong \tilde{H}_{n-1}(\partial\overline{X}_s)$. We put

$V_0 := \mathrm{Ker}(j)$, denote its rank by μ_0 and set $V' := V/V_0$. Then V' inherits from V a form which is nondegenerate. Clearly, Γ acts on V'. We let $\Gamma_s \subset \mathrm{Aut}(V')$ denote the image and $\Gamma_u \subset \Gamma$ the kernel of this representation. We have $\Gamma_s = \Gamma/\Gamma_u$, of course, but as W. Janssen pointed out, Γ is not always a semi-direct product of Γ_u and Γ_s. We shall call Γ_u resp. Γ_s the *unipotent* resp. *simple part* of Γ. This terminology is explained by the next two propositions.

(7.10) *Proposition.* There are no nontrivial Γ-invariant subspaces of $V'_{\mathbb{Q}}$

Proof. We show that any Γ-invariant subspace F of $V_{\mathbb{Q}}$ is either $V_{\mathbb{Q}}$ or contained in $(V_0)_{\mathbb{Q}}$. If $\delta \in \Delta$ is not orthogonal to F, then there exists an $x \in F$ with $(\delta.x) \neq 0$. Since $\sigma_\delta(x) = x \pm (x.\delta)\delta \in F$, it follows that $\delta \in F$. As $\Delta = \Gamma.\{\delta,-\delta\}$, it follows that $\Delta \subset F$ and so we must have $V = F$.

(7.11) *Proposition.* The unipotent part Γ_u is abelian and via the exponential map

$$\exp : V' \otimes V_0 \to \mathrm{Aut}(V),\ \exp(v' \otimes v_0)(x) = x + (v'.x)v_0$$

isomorphic to a sublattice of $V' \otimes V_0$.

Proof. Let $v_1,\dots,v_\mu$ be a basis of V such that $v_1,\dots,v_{\mu_0}$ is a basis for V_0. Any automorphism of V which acts trivially on V_0 and V/V_0 is of the form $x \mapsto x + \phi_1(x)v_1 +\dots+ \phi_{\mu_0}(x)v_{\mu_0}$ for certain (unique) $\phi_1,\dots,\phi_{\mu_0} \in V'^*$. On the other hand, any element of Γ is of the form $x \mapsto x + (a_1.x)v_1 +\dots+ (a_\mu.x)v_\mu$ for certain $a_1,\dots,a_\mu \in V$. It follows that any $\sigma \in \Gamma_u$ is of the form $x \to x + (a_1.x)v_1 +\dots+ (a_{\mu_0}.x)v_{\mu_0}$ for certain unique $a_1,\dots,a_{\mu_0} \in V'$. The map $\log : \Gamma_u \to V' \otimes V_0$, $\sigma \to a_1 \otimes v_1 +\dots+ a_{\mu_0} \otimes v_{\mu_0}$ is easily checked to be an injective group homomorphism.

(7.12) It is likely that $\log(\Gamma_u)$ is always of finite index in $V' \otimes V_0$. What else do we know about Γ? It is clearly contained in the group G of automorphisms of V which respect $(\, . \,)$ and act trivially on V^*/V'. Janssen (1983), generalizing work of A'Campo, Wajnryb (1980) and Chmutov (1982), has shown that for any odd-dimensional icis Γ contains the group $G^{(2)}$ of $g \in \mathrm{Aut}(V)$ which act trivially on $V^*/2V'$ so that in particular, Γ is of finite index in G. When n is even, $(\, . \,)$ will be symmetric. Let μ_+ resp. μ_- denote the dimension of a maximal pos. resp. neg. definite subspace of $V_{\mathbf{R}}$ and let μ_0 denote the rank of V_0. Put $\varepsilon = (-1)^{\frac{1}{2}n}$. Then the collection of *oriented* maximal ε-definite subspaces forms an open subset Ω of the Grassmannian $G^{or}_{\mu_\varepsilon}(V_{\mathbf{R}})$. It has two connected components when $(\, . \,)$ is not definite. Any reflection with respect to $\delta \in \Delta$ leaves a point of Ω invariant (any $\alpha \in \Omega$ orthogonal to δ will do) and hence leaves each component of Ω invariant. So if we let $G^{\#} \subset G$ denote the subgroup (of index ≤ 2) of $g \in G$ which preserve either component, then $\Gamma \subset G^{\#}$. Ebeling (1981, 1983) has shown that in many cases $\Gamma = G^{\#}$. Moreover, in these cases Δ is precisely the set of $\delta \in V$ with $(\delta.\delta) = 2\varepsilon$ for which there exists a $v \in V$ with $(\delta.v) = 1$. It is not true that Γ is always of finite index in G (for most of the cusp singularities, $|G/\Gamma| = \infty$) but it is conjectured that if $\mu_{-\varepsilon} \neq 1$, $\Gamma = G^{\#}$ (the cusps of dim 2 have $\mu_{-\varepsilon} = 1$).

Back to our map $f : \overline{X} \to S$. Let $s_1 \in D$ be arbitrary. Choose for any $x \in C_{s_1}$ a closed ball B_x in X centered at x (with $B_x \cap B_{x'} = \emptyset$ if $x \neq x'$) and a spherical neighbourhood S_1 of s_1 in S such that for all $x \in C_{s_1}$, $f : X_{S_1} \cap B_x \to S_1$ is a good representative for the germ of f at x. We choose $s_2 \in S_1 - (S_1 \cap D)$ and put $\overline{Y}_x := X_{s_2} \cap B_x$.

(7.13) *Proposition*. There is a natural exact sequence of *free* $\mathbf{Z}$-modules

$$0 \to \bigoplus_{x \in C_{s_1}} \tilde{H}_n(\overline{Y}_x) \overset{i_*}{\to} \tilde{H}_n(\overline{X}_{s_2}) \to \tilde{H}_n(\overline{X}_{s_1}) \to 0$$

where i_* is induced by inclusion (so that $\oplus_x \tilde{H}_n(\overline{Y}_x)$ can be identified with a primitive submodule of $\tilde{H}_n(\overline{X}_{s_2})$). The map i_* respects the intersection forms and sends vanishing cycles to vanishing cycles. In fact, if $\delta'_1,\ldots,\delta'_{m'} \in \cup_x \tilde{H}_n(\overline{Y}_x)$ is a distinguished system of generators for $\oplus_x \tilde{H}_n(\overline{Y}_x)$, then $\delta_1 := i_*(\delta'_1),\ldots,\delta_{m'} := i_*(\delta'_{m'})$ extends to a distinguished system of generators $\delta_1,\ldots,\delta_m$ for $\tilde{H}_n(\overline{X}_{s_2})$ where m resp. m' is the multiplicity of D at 0 resp. s_1).

(A submodule M of a free **Z**-module N is called *primitive* if N/M is torsion free.)

Proof. Since $f : \overline{X}_{s_1} - \cup_x(\mathring{B}_x \cap \overline{X}_{s_1}) \to S_1$ is a trivial fibre bundle we have natural isomorphisms

$$H_p(\overline{X}_{s_2}, \cup_x \overline{Y}_x) \overset{\cong}{\to} H_p(\overline{X}_{S_1}, \cup_x(\overline{X}_{S_1} \cap B_x)) \overset{\cong}{\leftarrow} H_p(\overline{X}_{s_1}, \cup_x(\overline{X}_{s_1} \cap B_x)).$$

Each $\overline{X}_{s_1} \cap B_x$ is contractible, so the last term may be identified with $H_p(\overline{X}_{s_1}, C_{s_1})$. As $\overline{X}_{s_1}$ has the homotopy type of a bouquet of n-spheres, it follows that $H_{n+1}(\overline{X}_{s_1}, C_{s_1}) = 0$ and that

$$0 \to \tilde{H}_n(\overline{X}_{s_1}) \to H_n(\overline{X}_{s_1}, C_{s_1}) \to \tilde{H}_{n-1}(C_{s_1}) \to 0$$

is exact. Feeding this into the exact reduced homology sequence

$$H_{n+1}(\overline{X}_{s_1}, C_{s_1}) \to \tilde{H}_n(\cup_x \overline{Y}_x) \to \tilde{H}_n(\overline{X}_{s_2}) \to H_n(\overline{X}_{s_1}, C_{s_1}) \overset{\partial_*}{\to} \tilde{H}_{n-1}(\cup \overline{Y}_x)$$

and using the fact that each $\overline{Y}_x$ is (n-1)-connected (so that $\tilde{H}_{n-1}(\cup \overline{Y}_x) \cong \tilde{H}_{n-1}(C_{s_1})$) gives the short exact sequence of the proposition. Each term is free since the corresponding space is homotopy equivalent to a bouquet of n-spheres.

Clearly, i_* respects the vanishing cycles and the

intersection form. The last assertion of the proposition is obtained by considering the intersection L of S with the line through s_1 parallel to the last coordinate axis: it meets D in m points (counted with multiplicity) and s_1 is a point of $D \cap L$ of multiplicity m'. Perturb L slightly to get an L' which intersects D in m regular points, exactly m' of which are in S_1. Moreover this set of m' points decomposes into subsets labeled by the points of C_{s_1}. Give the points of C_{s_1} a total order and do the same with each of the $|C_{s_1}|$ subsets of $D \cap L' \cap S_1$. Give the latter the lexicographical order and extend this to a total order of $D \cap L'$ (such that $t, t' \in D \cap L'$, $t \in S_1$, $t' \notin S_1$ implies $t < t'$). Then the construction of (7.4) gives a geometric system of generators as desired.

(7.14) It is not true in general that the group generated by the Picard-Lefschetz pseudo-reflections $\sigma_{i_*(\delta'_1)}, \ldots, \sigma_{i_*(\delta'_{m'})}$ restricts isomorphically to the group generated by $\sigma_{\delta_1}, \ldots, \sigma_{\delta_{m'}}$. In order to understand what happens, we take a closer look at the variation homomorphism, as defined in chapter 3.

Let $f : \overline{X} \to S$ be a good proper representative and fix some $s \in S-D$. As before we write V for $H_n(\overline{X}_s)$ and $V_0 \subset V$ for the radical of the intersection pairing on V. Poincaré duality enables us to identify the $\mathbb{Z}$-dual V* of V with $H_n(\overline{X}_s, \partial\overline{X}_s)$. So if $j : V \to V^*$ corresponds to the natural map $H_n(\overline{X}_s) \to H_n(\overline{X}_s, \partial\overline{X}_s)$, then $V_0 = \mathrm{Ker}(j)$ and $j^* = (-1)^n j$. We further write Π for $\pi(S-D,s)$. Given $\alpha \in \Pi$, the corresponding monodromy transformations $\rho_\alpha \in \mathrm{Aut}(V)$, $\rho_\alpha^{*-1} \in \mathrm{Aut}(V^*)$ can be written

(7.14.a) $$\rho_\alpha = \mathbf{1}_V + \mathrm{var}_\alpha \circ j \text{ and } \rho_\alpha^{*-1} = \mathbf{1}_{V^*} + j \circ \mathrm{var}_\alpha$$

respectively, where $\mathrm{var}_\alpha \in \mathrm{Hom}(V^*, V)$ denotes the variation along α (see

(3.1)). For $\alpha,\beta \in \Pi$ we find, by evaluating $var_{\alpha\beta}$ on a representative relative cycle:

(7.14.b) $$var_{\alpha\beta} = var_\alpha + var_\beta + var_\alpha \circ j \circ var_\beta.$$

Let $Var_\alpha \in V \otimes V$ correspond to var_α under the identification $Hom(V^*,V) \cong V \otimes V$. (In general, if V_1 and V_2 are $\mathbf{Z}$-modules, V_2 assumed to be free and of finite type, then we may identify $Hom(V_1,V_2)$ with $V_1^* \otimes V_2$.) Now (7.14.b) may be written as

(7.14.c) $$Var_{\alpha\beta} = (\mathbf{1} \otimes \rho_\alpha)Var_\beta + Var_\alpha.$$

(In other words, Var is a 1-cocycle of Π with values in the representation $\mathbf{1} \otimes \rho$ on $V \otimes V$.) This enables us to define a representation $\tilde{\rho}$ on $\mathbf{Z} \oplus (V \otimes V)$ by

$$\tilde{\rho}_\alpha(\lambda, v_1 \otimes v_2) = (\lambda, \lambda Var_\alpha + v_1 \otimes \rho_\alpha(v_2)).$$

It is clear that $\tilde{\rho}$ sits in a short exact sequence of representations:

(7.14.d) $$\underset{\mathbf{1}\otimes\rho}{0 \to V \otimes V} \to \underset{\tilde{\rho}}{\mathbf{Z} \oplus (V \otimes V)} \to \underset{\mathbf{1}}{\mathbf{Z}} \to 0.$$

Let $\tilde{\Gamma} \subset Aut(\mathbf{Z} \oplus (V \otimes V))$ denote the image of this representation. It is clear that $\tilde{\Gamma}$ projects in $Aut(V \otimes V)$ onto $\mathbf{1} \otimes \Gamma \cong \Gamma$. Before we say anything about the kernel of $\tilde{\Gamma} \to \Gamma$, we prove a universal property of $\tilde{\Gamma}$.

Consider the map $E : \mathbf{Z} \oplus (V \otimes V) \to V^* \otimes V \cong End(V)$, defined by

$$E(\lambda, v_1 \otimes v_2) := \lambda \mathbf{1}_V + j(v_1) \otimes v_2.$$

If we give $V^* \otimes V$ the representation $\mathbf{1} \otimes \rho$, then E becomes equivariant:

$$\begin{aligned}E\circ\tilde{\rho}_\alpha(\lambda,v_1\otimes v_2) &= \lambda\mathbf{1}_V + \lambda(j\otimes\mathbf{1})\,\mathrm{Var}_\alpha + j(v_1)\otimes\rho_\alpha(v_2)\\ &= \lambda\mathbf{1}_V + \lambda(\rho_\alpha-\mathbf{1}_V) + j(v_1)\otimes\rho_\alpha(v_2)\\ &= \mathbf{1}\otimes\rho_\alpha(\lambda\mathbf{1}_V + j(v_1)\otimes v_2)\\ &= (\mathbf{1}\otimes\rho_\alpha)E(\lambda,v_1\otimes v_2).\end{aligned}$$

More generally, suppose we are given a free $\mathbb{Z}$-module W of finite type endowed with a bilinear form (.) and a homomorphism $i : V \to W$ which respects the forms: $(i(v_1).i(v_2)) = (v_1.v_2)$. So if $k : W \to W^*$ denotes the adjoint of the form $(k(w_1)(w_2) = (w_1.w_2))$, then $i^*ki = j$. It is easy to verify that $\rho^i_\alpha(w) := w + i\,\mathrm{var}_\alpha i^*k(w)$ defines a representation of Π on W which makes i equivariant. (The example to think of is when W is the Milnor lattice of a singularity which has (X_0,x) in its miniversal deformation.) We claim that the image Γ^i of this representation is in a natural way a quotient of $\tilde{\Gamma}$. To see this, define $E^i : \mathbb{Z}\oplus(V\otimes V) \to W^*\otimes W \cong \mathrm{End}(W)$ by $E^i(\lambda,v_1\otimes v_2) := \lambda\mathbf{1}_W + ki(v_1)\otimes i(v_2)$. As before, E^i becomes equivariant if we give $W^*\otimes W$ the representation $\mathbf{1}\otimes\rho^i$. In particular $E^i\circ\tilde{\rho}_\alpha(1,0) = (\mathbf{1}\otimes\rho^i_\alpha)\circ E^i(1,0) = (\mathbf{1}\otimes\rho^i_\alpha)\mathbf{1}_W = \rho^i_\alpha$ so that E^i induces a surjection $\tilde{\Gamma}\to\Gamma^i$, $\tilde{\rho}_\alpha \mapsto \rho^i_\alpha$. If $ki\otimes i : V\otimes V \to W^*\otimes W$ is injective, then it is easily seen that E^i is injective, unless W is of rank 1. This implies that $\tilde{\Gamma}\to\Gamma^i$ is then an isomorphism (the case $\mathrm{rk}(W) = 1$ is rather trivial). Clearly, $ki\otimes i$ is injective if and only if ki is and this condition is satisfied for instance if i is injective and (W,(.)) is nondegenerate. So we can always realize $\tilde{\Gamma}$ as some Γ^i. In this context we mention (without proof) that (X_0,x) is always in the miniversal deformation of a singularity whose intersection form is nondegenerate.

Next we have a look at the kernel Z of $\tilde{\Gamma}\to\Gamma$. We begin with deriving the following formula

(7.14.e) $\quad (-1)^n\mathrm{var}_\alpha + \mathrm{var}^*_\alpha + \mathrm{var}^*_\alpha\circ j\circ\mathrm{var}_\alpha = 0.$

Let C and C' be relative n-cycles on $(\overline{X}_s,\partial\overline{X}_s)$ in general position (i.e. C and C' meet in X_s only and there they intersect transversally) and let $h : (\overline{X}_s,\partial\overline{X}_s)\circlearrowleft$ represent the C^∞-monodromy along $\alpha \in \Pi$. Then $h(C).C' - C . C' = h(C).C' - h(C).h(C')$. Passing to homology classes: $c := [C]$, $c' := [C']$, yields $(h(C) - C).C' = (-1)^n < c',var_\alpha(c)>$ and $h(C).(C' - h(C')) = <c + jvar_\alpha(c),-var_\alpha(c')>$. Equating these gives (7.14.e).

Now suppose that $\rho_\alpha = \mathbf{1}$. The exact sequence (7.14.d) shows that then $\tilde{\rho}_\alpha$ is of the form $(\lambda,v_1 \otimes v_2) \to (\lambda, v_1 \otimes v_2 + \lambda Var_\alpha)$. The properties $j\circ var_\alpha = 0$, $var_\alpha\circ j = 0$ show that in fact $Var_\alpha \in V_0 \otimes V_0$. If $\beta \in \Pi$ is arbitrary, then it is easily verified that $\tilde{\rho}_\beta\circ\tilde{\rho}_\alpha = \tilde{\rho}_\alpha\circ\tilde{\rho}_\beta$. So Z is central in $\tilde{\Gamma}$ and $\tilde{\rho}_\alpha \to Var_\alpha$ identifies Z with an (additive) subgroup Z of $V_0 \otimes V_0$. It follows from (7.14.e) that Z consists of symmetric resp. antisymmetric integral bilinear forms if n is odd resp. even (so has the opposite symmetry of (.)).

We shall refer to $\tilde{\Gamma}$ as the *variation extension* of Γ.

7.C *Adjacency*

(7.15) We are going to define a partial ordering on the set $\mathcal{Cis}(n)$ of isomorphism classes of non-regular icis's of dim n. This will measure how complex an icis is compared to other icis's of the same dimension. The isomorphism class of an icis (X_0,x) of dim n will be denoted by $[(X_0,x)]$.

In order to define the partial ordering, take any icis (X_0,x) of dim n. Choose a versal deformation $f : (\mathbb{C}^{n+k},x) \to (\mathbb{C}^k,0)$ of (X_0,x) and let $f : X \to \mathbb{C}^k$ be a representative which is versal in each point of X.

Given an isomorphism class $[(Y_0,y)] \in \mathcal{Cis}(n)$, we say that $[(Y_0,y)]$ is *adjacent* to $[(X_0,x)]$ or that $[(X_0,x)]$ *deforms* to $[(Y_0,y)]$ - and we write $[(X_0,x)] \to [(Y_0,y)]$ - if x is in the closure of the set of $z \in X$ with $(X_{f(z)},z)$ isomorphic to (Y_0,y). It is clear that this is really a relation between isomorphism classes (in particular it is independent of the choice of f) and that this relation is transitive. Without proof we mention that the set of $z \in X$ with $(X_{f(z)},x)$ isomorphic to (X_0,x) is closed in X (compare the discussion in (6.15)). This implies that adjacency is a partial ordering on $\mathcal{Cis}(n)$.

We define the *modality* $\mathrm{mod}[(X_0,x)]$ of $[(X_0,x)]$ as the smallest integer m for which there exists a representative $f : X \to \mathbb{C}^k$ of a versal deformation of (X_0,x) and an analytic subset of X of $\dim \le m$ which meets every equivalence class in X. Then $[(X_0,x)]$ is called m-*modal*. For $m = 0,1,2$ it is customary to say that the class is *simple*, *unimodal*, *bimodal* respectively. So $[(X_0,x)]$ is simple means that (X_0,x) admits a versal deformation in which only a finite number of isomorphism types of singularities occur. Our notions of adjacency and modality are not the same as Arnol'd's (1974): he reserves these notions for germs of functions and takes "right-equivalence" instead of "isomorphism" as the fundamental relation. If (X_0,x) deforms to (Y_0,y) then the invariants of (Y_0,y) tend to be simpler c.q. smaller than those of (X_0,x), for instance

(i) $\mathrm{mod}[(Y_0,y)] \le \mathrm{mod}[(X_0,x)]$ (clear),

(ii) $\tau(Y_0,y) \le \tau(X_0,x)$, by (6.15.1),

(iii) $\mathrm{emb.cod}(Y_0,y) \le \mathrm{emb.cod}(X_0,x)$ (clear),

(iv) the vanishing lattice of (Y_0,y) embeds primitively in the vanishing lattice of (X_0,x) in the way described by (7.13), in particular $\mu(Y_0,y) \le \mu(X_0,x)$.

It can happen (we will see an example of this in Section D) that for a family $\{(Y_t,y_t)\}_{t\in T}$ of icis's, none of the individual (Y_t,y_t) is adjacent to (X_0,x) while x is nevertheless in the closure of the union of the equivalence classes in X corresponding to the (Y_t,y_t). For this reason one often collects isomorphism classes which are similar (have the same topological type, Milnor number etc. and are incomparable with respect to the adjacency relation) and extends the notion of adjacency to such subsets of $\mathcal{Cis}(n)$.

Every icis deforms to a quadratic singularity (by (4.10) and (6.11)), unless it is regular, of course. So the isomorphism class of a quadratic singularity of dim n (which we denote by A_1) is the unique minimal element of $\mathcal{Cis}(n)$. The classification of isomorphism classes of icis's of dim n is in principle accomplished by working our way up from A_1 in the partially ordered set $\mathcal{Cis}(n)$. In the beginning we only find hypersurface singularities. In fact, it appears that for large n, we must be quite high up before we leave the domain $\mathcal{Hyp}(n) \subset \mathcal{Cis}(n)$ of isomorphism classes of isolated hypersurface singularities, for the singularity described in example 1 of Ch. 5 represents the unique minimal element of $\mathcal{Cis}(n) - \mathcal{Hyp}(n)$ and has modality $\geq$ n-1. An interesting aspect of the hypersurface classification is that it is not independent of n, as will become clear in (7.17) below. Define the *hessian corank* of $f \in m^2_{\mathbb{C}^{n+1},0}$ as the dimension of the nilspace of the matrix $(\frac{\partial^2 f}{\partial z_\nu \partial z_\mu}(0))$. It is easily checked that this is an invariant of the local $\mathbb{C}$-algebra $\mathcal{O}_{\mathbb{C}^{n+1},0}/f\mathcal{O}_{\mathbb{C}^{n+1},0}$, so that it makes sense to speak of the hessian corank of an element of $\mathcal{Hyp}(n)$. The following lemma is due to Gromoll & Meyer (1969).

(7.16) *Splitting lemma.* If $f \in m^2_{\mathbb{C}^{n+1},0}$ has hessian corank c, then there exists an automorphism $h : (\mathbb{C}^{n+1},0) \circlearrowleft$ such that

$$f\circ h(z) = f'(z_1,\dots,z_c) + z_{c+1}^2 + \dots + z_{n+1}^2$$

with $f' \in m^3_{\mathbb{C}^c,0}$. Moreover, f' is unique up to source equivalence: if $f'' \in m^3_{\mathbb{C}^c,0}$ is obtained in the same way then there exists an automorphism $h' : (\mathbb{C}^c,0)\circlearrowleft$ such that $f'' = f'\circ h'$.

Proof. By making a linear coordinate change in $(\mathbb{C}^{n+1},0)$ we may achieve that the restriction of f to the subspace defined by $z_1 = \dots = z_c = 0$ has a nondegenerate hessian. Write $z = (z',z'') \in \mathbb{C}^c\times\mathbb{C}^{n+1-c}$ so that $z'' \to f(0,z'')$ is nondegenerate. The Morse lemma with parameters $z_1,\dots,z_c$ shows that there is an automorphism $h : (\mathbb{C}^{n+1},0)\circlearrowleft$ which converts f into the above form. The various f' we obtain are all gotten by restricting f to a nonsingular subgerm of $(\mathbb{C}^{n+1},0)$ of dim c defined by $(n+1-c)$ elements of the ideal $(\frac{\partial f}{\partial z_1},\dots,\frac{\partial f}{\partial z_{n+1}})\mathcal{O}_{\mathbb{C}^{n+1},0}$ with *independent* differentials in 0. The collection of such $(n+1-c)$-tuples is a connected set and an argument similar to the one employed in the proof of (6.3) will show that f' is unique up to source equivalence (we omit the details).

(7.17) We call f' the *reduced form* of f. It is easy to check that if $f' : (\mathbb{C}^c,0) \to (\mathbb{C},0)$ defines an isolated hypersurface singularity the addition of squares in new variables: $f(z) = f'(z') + z_{c+1}^2 + \dots + z_{n+1}^2$ does not alter its Tjurina number (nor its Milnor number). In fact, a miniversal deformation $F' : (\mathbb{C}^c\times\mathbb{C}^{\tau-1},0) \to (\mathbb{C}\times\mathbb{C}^{\tau-1},0)$ of $(f'^{-1}(0),0)$ of the form $F'(z',u) = (g(z',u),u)$ determines one of $(f^{-1}(0),0)$ by taking $F(z,u) = (g(z',u) + z_{c+1}^2 + \dots + z_{n+1}^2, u)$. If we combine this with the splitting lemma we find that the procedure of adding (or removing) a fixed number of squares in separate variables does not affect the adjacency relation. In particular, the partially ordered subsets of $Hyp(n)$ and $Hyp(m)$ defined by the condition that the hessian corank is $\leq \min(n,m)$ are naturally isomorphic. So when passing from $Hyp(n-1)$ to

$Hyp(n)$ we only need to worry about singularities of multiplicity ≥ 3. The modality of such a singularity can be shown to be $\geq (n^3-n)/6$.

Another application of (7.17) is

(7.18) *Proposition.* An icis $(X_0,0)$ of dim n which is neither regular nor a quadratic singularity deforms to the isomorphism class A_2 defined by $z_1^3 + z_2^2 + \ldots + z_{n+1}^2$.

Proof. Let us first assume that $(X_0,0)$ is a hypersurface singularity. Because of the preceding discussion it suffices to treat the case when the defining equation f is in $m^3_{\mathbf{C}^{n+1},0}$. We do this with the help of the deformation $(z,u) \to (f(z) + u(z_2^2 + \ldots + z_{n+1}^2), u)$. For small $u \neq 0$, there is a source automorphism which converts f_u into a function of the form $g(z_1) + z_2^2 + \ldots + z_{n+1}^2$ with $g \in m^3_{\mathbf{C},0}$. If v is a small number $\neq 0$, then the cubic root of $g(z_1) + vz_1^3$ can be extracted and hence $g(z_1) + vz_1^3$ will be source equivalent to z_1^3. This settles the hypersurface case.

If $(X_0,0)$ is not a hypersurface germ, then it can be defined by a map-germ $f : (\mathbf{C}^{n+k},0) \to (\mathbf{C}^k,0)$ with $df(0) = 0$ and $k \geq 2$. Denote the quadratic part (= 2-jet) of f_κ by q_κ. Some nontrivial linear combination of q_1 and q_2 is degenerate (this is true for any pair of quadratic forms), so by making a linear coordinate change in $\mathbf{C}^k$ we may as well assume that q_1 is degenerate. Another coordinate change in $\mathbf{C}^{n+k}$ will make q_1 independent of z_1. Consider the deformation $(z,u_2,\ldots,u_k) \to (f_1(z), f_2(z) + u_2z_2, \ldots, f_k(z) + u_kz_k, u_2, \ldots, u_k)$. For almost all u, f_u will define a hypersurface germ at 0 which is not a quadratic singularity. These deform to an A_2-singularity, hence so does $(X_0,0)$.

7.D *A partial classification*

We give a brief outline of the beginning of the classification. For proofs, c.q. a description of the method by which the classification is brought about, the reader should consult the references given in the text. We shall describe successively

(a) the simple singularities in dim 0

(b) the simple hypersurface singularities (their hessian corank appears to be always 1 or 2)

(c) the simple singularities not belonging to the previous classes (they turn out to be space curve singularities: dim = 1, emb.dim = 3)

(d) some of the unimodal germs in dim 2.

In the cases (b) and (d) we shall also describe the vanishing lattices. Occasionally we will comment on the (possible) adjacency relations.

(7.19) *Simpleness in dim 0.* The classification of simple 0-dim icis's is due to Giusti (1977). We use his notation, in which the lower index denotes the Milnor number. (By (5.12) this is one less than the degree of a defining map.) The cited reference gives many (and presumably all) adjacency relations among them.

symbol	equations	Tjurina number
A_μ, $1 \le \mu$	$(x^{\mu+1})$	μ
$F^{p,q}_{p+q-1}$, $2 \le p \le q$	(xy, x^p+y^q)	$p+q$
G_5	(x^2, y^3)	7
G_7	(x^2, y^4)	10
H_μ, $6 \le \mu$	$(x^2+y^{\mu-3}, xy^2)$	$\mu+2$

symbol	equations	Tjurina number
I_μ, $7 \le \mu$	$\left\{\begin{array}{l}(x^2+y^3,y^k) \text{ if } \mu=2k-1\\ (x^2+y^3,xy^{k-1}) \text{ if } \mu=2k\end{array}\right\}$	$\mu+2$

(7.20) In the case of a hypersurface data pertaining to a vanishing lattice can be conveniently codified in an *intersection diagram*. This presupposes that we are given a distinguished basis $\delta_1,\ldots,\delta_\mu$ of the vanishing lattice. The intersection diagram is then a graph on $\{\delta_1,\ldots,\delta_\mu\}$ of which each edge has been given a weight (and if n is odd, also an orientation): δ_i and δ_j $(i \neq j)$ are connected if and only if $(\delta_i.\delta_j) \neq 0$; if n is even we give the edge $\{\delta_i,\delta_j\}$ weight $(-1)^{\frac{1}{2}n}(\delta_i.\delta_j)$; if n is odd we give it weight $|(\delta_i.\delta_j)|$ and orient $\{\delta_i,\delta_j\}$ by the condition that $\delta_i > \delta_j$ if $(\delta_i.\delta_j) > 0$. It is clear that this diagram determines the vanishing lattice up to isomorphism (even if we forget the enumeration of the vertices). A disadvantage is that the intersection diagram is anything but unique. However in the cases we consider the diagram has a tendency to become unique if we insist that the number of cycles in the graph is as small as possible. If n is odd and the intersection diagram is a tree, then by replacing some of the $\delta_1,\ldots,\delta_\mu$ by their antipodes we can achieve an arbitrary orientation of the edges, so in such a case there is no need to specify the orientation. As many weights will be $(-1)^{n+1}$, we shall only indicate the weights $\neq (-1)^{n+1}$. For all the hypersurface singularities discussed here the intersection diagram has been determined by Gabrielov (1974).

(7.21) *Simple hypersurface singularities*. Their classification is due to Arnol'd. The result is quite surprising: it appears that up to a sum of squares in separated variables they are just the Kleinian singularities!

Besides, their intersection diagrams will be familiar to readers who know about Coxeter groups.

Symbol	equation	intersection diagram
A_μ, $1 \le \mu$	$z_1^{\mu+1}+z_2^2+\ldots+z_{n+1}^2$	
D_μ, $4 \le \mu$	$z_1^{\mu-1}+z_1z_2^2+z_3^2+\ldots+z_{n+1}^2$	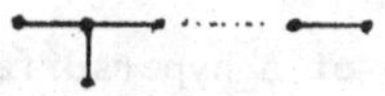
E_6	$z_1^3+z_2^4+z_3^2+\ldots+z_{n+1}^2$	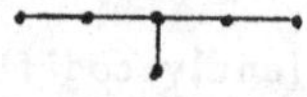
E_7	$z_1^3+z_1z_2^3+z_3^2+\ldots+z_{n+1}^2$	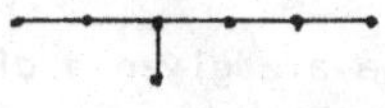
E_8	$z_1^3+z_2^5+z_3^2+\ldots+z_{n+1}^2$	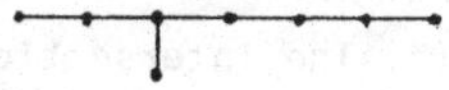

Again, the lower index of the symbol represents the Milnor number (which is also the Tjurina number). The adjacency relations among these singularities correspond exactly to the inclusion relations of their intersection diagrams. In the even-dimensional case, the form $(-1)^{\frac{1}{2}n}(\ .\ )$ is positive definite (this actually characterizes the simple singularities of positive even dimension) and so Γ is finite. The finite reflection groups have been classified a long time ago by Coxeter; those with the property that they leave a lattice invariant and have their reflections in a single conjugacy class are precisely the groups occurring here. In fact, more is true: Δ is a root system relative the form $(-1)^{\frac{1}{2}n}(\ .\ )$ and a distinguished basis with an intersection diagram as above is a root basis for Δ (Bourbaki (Lie, VI)). Behind this lies a more direct relationship between the Kleinian singularities and the simple Lie groups, see Brieskorn (1970b), Slodowy (1980), Gonzalez-Springberg & Verdier (1981) and Knörrer (1982b).

Other characterizations of Kleinian singularities can be found in a survey by Durfee (1979).

(7.22) *Remaining simple singularities*. These are all singularities of space curves. Most of them occur in Mather's work (1971) but their classification is due to Giusti (1977). We stick to his notation ($\mu = \tau$ is the lower index)

symbol	equations
S_μ, $5 \le \mu$	$(z_1^2+z_2^2+z_3^{\mu-3}, z_2z_3)$
T_μ, $\mu=7,8,9$	$(z_1^2+z_2^3+z_3^\mu, z_2z_3)$
U_7	$(z_1^2+z_2z_3, z_1z_2+z_3^3)$
U_8	$(z_1^2+z_2z_3, z_1z_2+z_1z_3^2)$
U_9	$(z_1^2+z_2z_3, z_1z_2+z_3^4)$
W_8	$(z_1^2+z_2z_3, z_2^2+z_3^3)$
W_9	$(z_1^2+z_2z_3, z_2^2+z_1z_3^2)$
Z_9	$(z_1^2+z_2^2+z_3^3, z_1z_2)$
Z_{10}	$(z_1^2+z_2z_3^2, z_2^2+z_3^3)$.

Giusti also describes their vanishing lattices and some, if not all, adjacency relations among these singularities and the singularities in the preceding list.

(7.23) *Some unimodal germs of dim 2.* The classification of 2-dimensional complete intersections begins with the Kleinian singularities. Beyond this range we enter the wonderland of unimodal germs. We meet successively

(α) the simply-elliptic singularities $\tilde{D}_5, \tilde{E}_6, \tilde{E}_7, \tilde{E}_8$ (Ch. 1, ex. 5)

(β) the cusp singularities $T_{p,q,r}$ and $T_{p,q,r,s}$ (Ch. 1, ex. 9)

(γ) the triangle singularities $D_{p,q,r}$ (Ch. 1, ex. 6) with their non-quasi-homogeneous companion $D'_{p,q,r}$ (Ch. 1, ex. 6 cont.) for precisely 22 triples (p,q,r).

These singularities are unimodal, although this list doesn't exhaust all 2-dimensional unimodal germs. In some instances we describe their vanishing lattices by means of intersection diagrams: in the non-hypersurface case the substitute for a geometric basis is just a basis of vanishing cycles with the property that the corresponding reflections generate the monodromy group. We will also say a few words on the abstract properties of the vanishing lattice.

(α) The *simply-elliptic singularities* of type $\tilde{D}_5, \tilde{E}_6, \tilde{E}_7, \tilde{E}_8$ (see Ch. 1, ex. 5 for their equations). An intersection diagram is given under (β) with $(p,q,r,s) = (2,2,2,2)$, $(3,3,2,1)$, $(2,4,3,1)$, $(2,3,5,1)$ respectively. The Milnor lattice V has a radical V_0 of rank 2 and $V' = V/V_0$ is isomorphic to the Milnor lattice of the Kleinian singularity D_5, E_6, E_7, E_8 respectively. So we have $\mu_0 = 2$, $\mu_+ = 0$ and μ_- is just the lower index. An element $\delta \in V$ is a vanishing cycle if and only if $(\delta.\delta) = -2$. We have $\Gamma_u = \exp(V'\otimes V_0)$ and Γ_s can be identified with the monodromy group (a Coxeter group) of D_5, E_6, E_7, E_8 respectively. Using this, it can be shown that $\Gamma = G = G^{\#}$. The simply-elliptic singularities have no internal adjacency relations: the (unique) maximal element of $\mathcal{C}is(2)$ dominated by some $\tilde{E}_8$-singularity is E_8 (and similarly for the others). The miniversal deformations of these singularities have been investigated by Pinkham (1977d) (see also Mérindol (1982)) and Looijenga (1977, 1978).

(β) The *cusp singularities* $T_{p,q,r}$ $(\frac{1}{p} + \frac{1}{q} + \frac{1}{r} < 1)$ and $T_{p,q,r,s}$ $((\frac{1}{p} + \frac{1}{q})(\frac{1}{r} + \frac{1}{s}) < 1)$. They have the following intersection diagram (recall that $T_{p,q,r,1} = T_{p,q,r+1}$)

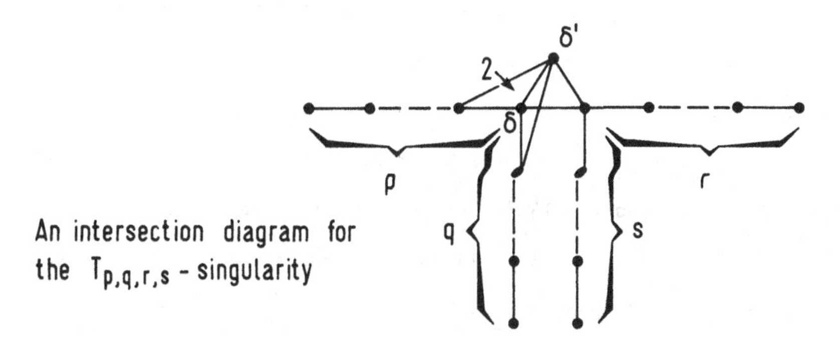

An intersection diagram for the $T_{p,q,r,s}$ - singularity

The radical V_0 of V is of rank one and spanned by $\delta-\delta'$, so that V' may be identified with the sublattice spanned by all basis vectors except δ'. The signature of V' is $(1,\mu-2)$. We have $\Gamma_u = \exp(V'\otimes V_0)$. The group Γ_s with its generating set of reflections $\sigma_{\delta''}$ (δ'' running over the basis, $\delta'' \neq \delta'$) is a Coxeter system in the sense of Bourbaki and its action on $V'_{\mathbb{R}}$ is just the standard representation of the Coxeter system (Bourbaki, Lie V). In fact, we have in V' naturally defined a generalized root system Δ' in the sense of Kac-Moody Lie-algebras and Δ is just the pre-image of Δ' in V. Except for the simplest cusps $T_{3,3,4}$, $T_{2,4,5}$, $T_{2,3,7}$, $T_{2,2,2,3}$, the group Γ will not be of finite index in G. The adjacency relations among the cusps (and simpler singularities) can be found in Karras (1977). Here we encounter the phenomenon alluded to in (7.15): no individual $\tilde{E}_6$-singularity is adjacent to $T_{3,3,4}$, but the whole family is, by virtue of the deformation $z_1^4+z_2^3+z_3^3+z_1z_2z_3+tz_1^3$. For more about the miniversal deformation of a cusp, see Looijenga (1981).

(γ) The *triangle singularities* $D_{p,q,r}$, $p \leq q \leq r$ and $\frac{1}{p}+\frac{1}{q}+\frac{1}{r} < 1$. As remarked in Ch. 1, for 14 triples the embedding codimension is 1 and for 8 (remaining) triples the embedding codimension is 2. We shall not give equations (for these we refer to Arnol'd (1974) and Pinkham (1977b)), but only tell which triples occur. They naturally fall into 3 classes:

(i) $3 \le p$, $p+q+r \le 13$ (< 13 for a hypersurface) (10 triples)

(ii) $p = 2$, $4 \le q$, $p+q+r \le 14$ (< 14 for a hypersurface) (8 triples)

(iii) $p = 2$, $q = 3$, $p+q+r \le 15$ (< 15 for a hypersurface) (4 triples)

and in all cases $\mu = 22-p-q-r$, $\mu_0 = 0$, $\mu_+ = 2$. Intersection diagrams have been given by Gabrielov (1974) and Pinkham (1977c). The latter also gave an a priori description of Γ, of which Brieskorn observed that this comes down to the statement $\Gamma = G^{\#}$. There is also a simple characterization of the set of vanishing cycles: $\delta \in V$ is a vanishing cycle if and only if $(\delta.\delta) = -2$ and $(\delta.v) = 1$ for some $v \in V$. The adjacency relations (internal and with other unimodal germs) have been determined by Brieskorn (1979). Pinkham (1977c) discovered that the miniversal deformation of a $D_{p,q,r}$-singularity is closely related to the deformation theory of certain K3 surfaces (for a partial survey, see Looijenga (1983)).

Any icis of dim 2 not belonging to one of the classes discussed here deforms to some triangle singularity. So what we have here is really a beginning of the classification. Beyond the triangle singularities one still finds interesting classes of singularities, some of which may deserve further study. Arnol'd has pushed the classification of hypersurface singularities quite far, see his 1976 paper for the state of affairs as of that date. The classification of all unimodal complete intersections of dim > 0 was recently completed by Wall (1983).

8 THE LOCAL GAUSS-MANIN CONNECTION

A classical result due to De Rham asserts that the singular cohomology of a differentiable manifold (with real coefficients) coincides with the cohomology of its complex of differential forms. In this chapter we show that something similar is true in the case of a deformation $f : (X,x) \to (\mathbb{C}^k,0)$ of an isolated singularity: the cohomology groups of the relative De Rham complex $\Omega^{\bullet}_{f,x}$ describe the cohomology of not just a single fibre, but of all fibres simultaneously of some good representative of f. The way in which the cohomology groups of the various Milnor fibres fit together is expressed by a flat connection on the cohomology groups $H^p(\Omega^{\bullet}_{f,x})$ (considered as $\mathcal{O}_{\mathbb{C}^k,0}$-modules) which has a pole along the discriminant of f. This is the so-called Gauss-Manin connection. It was introduced by Brieskorn (1970a) in order to study hypersurface singularities and since then the theory has been further developed by Greuel (1975, 1980), Hamm (1974), Malgrange (1974b), Saito (1973, 1983), Pham (1979) and others.

Let us briefly review each section. The first section is concerned with the definition and properties of the relative De Rham complex of a good representative, in particular we discuss (but, by way of exception, do not prove) the coherence of De Rham cohomology sheaves. Section B is about the Gauss-Manin connection; we prove that it is regular singular. The last section focusses on the complete intersection case. Among other things we find two formulas for the Milnor number of an icis, which differ from those given in Ch. 5. Most of the results of this section are due to Greuel.

8.A *De Rham cohomology of good representatives*

Given a mapping of spaces $f : X \to S$ and an abelian sheaf F on X, one has defined for $p \geq 0$ the higher direct image $R^p f_* F$ as the sheaf on S associated to $S \supset V \mapsto H^p(f^{-1}(V), F)$. For $p = 0$ this gives us the ordinary direct image $f_* F$ and indeed $R^p f_*$ is the pth right derived functor of f_*. We start with a preparatory result concerning higher direct images of a good representative.

(8.1) *Lemma*. Let $f : X \to S$ be a good representative of a deformation of an isolated singularity. Then the natural map

$$(R^p f_*(\mathbb{C}_X) \otimes_{\mathbb{C}} \mathcal{O}_S \to R^p f_* f^{-1} \mathcal{O}_S$$

is an isomorphism.

For the proof we need another lemma (taken from Deligne (1970)).

(8.2) *Lemma*. Let A be a paracompact, locally contractible space whose Betti numbers are finite. Then for any analytic space S the natural map $H^p(A,\mathbb{C}) \otimes_{\mathbb{C}} \mathcal{O}_S \to R^p \pi_* \pi^{-1} \mathcal{O}_S$ is an isomorphism. Here $\pi : A \times S \to S$ is the projection.

Proof. It is enough to show that for any compact Stein subset $V \subset S$ the natural map $H^p(A,\mathbb{C}) \otimes_{\mathbb{C}} H^0(V,\mathcal{O}_S) \to H^p(A \times V, \pi^{-1}\mathcal{O}_S)$ is an isomorphism as such

V generate the topology of S. For this, we consider the Leray sequence of the other projection $\pi_A : A\times V \to A$ (Godement (1958), II,4.17.1 and the subsequent remark):

$$E_2^{pq} = H^p(A,H^q(V,\mathcal{O}_S)) \Rightarrow H^{p+q}(A\times V,\pi^{-1}\mathcal{O}_S)$$

Since $H^q(V,\mathcal{O}_S) = 0$ for $q > 0$, the sequence collapses:

$$H^p(A,H^0(V,\mathcal{O}_S)) \overset{\cong}{\to} H^p(A\times V,\pi^{-1}\mathcal{O}_S)$$

The fact that A has finite Betti numbers permits us to invoke the universal coefficient theorem (Spanier (1966), 6.8.10) which says that the pairing $H^p(A,\mathbb{C}) \otimes_{\mathbb{C}} H^0(V,\mathcal{O}_S) \to H^p(A,H^0(V,\mathcal{O}_S))$ is an isomorphism.

Proof of (8.1). For any subset $Z \subset X$, let ϕ_Z denote the natural map $(R^pf_*\mathbb{C}_Z) \otimes_{\mathbb{C}} \mathcal{O}_S \to R^pf_*(f^{-1}\mathcal{O}_S|Z)$. We choose an open neighbourhood N of $\partial\overline{X}$ in $\overline{X}$ such that the pair $(N,X\cap N)$ is topologically trivial over S. Then the preceding lemma implies that ϕ_N and $\phi_{N\cap X}$ are isomorphisms. Since $f|\overline{X}$ is proper, it follows that

$$(R^pf_*f^{-1}\mathcal{O}_S)_s = \lim_{V\ni s} H^p(\overline{X}_V,f^{-1}\mathcal{O}_S) \cong H^p(\overline{X}_s,\mathcal{O}_{S,s})$$

(Godement (1958), II,4.11.1) and by the universal coefficient formula, the latter is just $H^p(\overline{X}_s,\mathbb{C}) \otimes_{\mathbb{C}} \mathcal{O}_{S,s}$. So $\phi_{\overline{X}}$ is also an isomorphism. A Mayer-Vietoris argument applied to the pair (N,X) (or better: the spectral sequence corresponding to this covering) implies that ϕ_X is also an isomorphism.

(8.3) One says that a morphism $f : X \to S$ of analytic spaces is a *Stein morphism* if for any Stein-open subset V of S, $f^{-1}(V)$ is a Stein-open subset of X. Such maps have the property that for any coherent $\mathcal{O}_X$-module

F, the higher direct images R^pf_*F, $p \geq 1$ vanish. This implies that the f-direct image of an exact sequence of coherent $\mathcal{O}_X$-modules is still exact. If we are given a deformation $f : (X,x) \to (\mathbb{C}^k,0)$ of an isolated singularity, then a simple way to obtain good Stein representatives for f is by taking the function r used to define x in $f^{-1}(0)$ (see (2.3)) of the form $|\phi_1|^2+\ldots+|\phi_\ell|^2$ where $\phi_1,\ldots,\phi_\ell \in m_{X,x}$ are such that their common zero set intersects (X_0,x) in $\{x\}$ only. For $g = (f,\phi_1,\ldots,\phi_\ell) : (X,x) \to (\mathbb{C}^{k+\ell},0)$ is then a finite germ and if we choose a good representative of g of the form $X \to S{\times}B$ ($S \subset \mathbb{C}^k$ open, $B \subset \mathbb{C}^\ell$ an open ball), then for any Stein-open $V \subset S$, $V{\times}B$ is Stein and hence $g^{-1}(V{\times}B) = f^{-1}(V)$ is Stein (for g is finite).

(8.4) Now recall from 6.A that associated to a morphism of analytic spaces we have a universal $f^{-1}\mathcal{O}_S$-derivation $d_f : \mathcal{O}_X \to \Omega_f$. This derivation extends uniquely to a square zero $f^{-1}\mathcal{O}_S$-derivation of the $\mathcal{O}_X$-exterior algebra of Ω_f (see e.g. Bourbaki: Alg. X). We write Ω_f^p for $\Lambda^p_{\mathcal{O}_X}\Omega_f$ (with the usual convention that $\Omega_f^0 = \mathcal{O}_X$) and refer to it as the sheaf of *holomorphic p-forms relative* f. So we have a complex of $f^{-1}\mathcal{O}_S$-modules

$$\Omega_f^{\cdot} : 0 \to \Omega_f^0 \xrightarrow{d_f^0} \Omega_f^1 \xrightarrow{d_f^1} \Omega_f^2 \to \ldots,$$

called the *De Rham complex* of f. The cohomology sheaf $\mathcal{H}^p(\Omega_f^{\cdot})$ (i.e. the sheaf on X associated to $X \supset U \mapsto H^p(\Gamma(U,\Omega_f^{\cdot}))$) will clearly be a sheaf of $f^{-1}\mathcal{O}_S$-modules. If f is a submersion in $z \in X$, then a simple parametrized version of the Poincaré lemma (see for instance Deligne (1970), I,2.23.2) asserts that $H^p(\Omega_{f,z}^{\cdot}) = 0$ for $p > 0$ and $H^0(\Omega_{f,z}^{\cdot}) \cong \mathcal{O}_{S,f(z)}$; in other words, $\Omega_{f,z}^{\cdot}$ is a resolution of $(f^{-1}\mathcal{O}_S)_z$. We can also form the more global object $\mathcal{H}^p(f_*\Omega_f^{\cdot})$; this is the sheaf on S associated to $S \supset V \mapsto H^p(\Gamma(f^{-1}(V),\Omega_f^{\cdot}))$. It is an $\mathcal{O}_S$-module, called the pth *De Rham* *cohomology sheaf* of f. Its formation is compatible with base change.

As is well known, holomorphic p-forms on an analytic manifold X can be integrated over C^1 p-chains on X. This remains true if X has singularities, for holomorphic forms always extend locally in an ambient nonsingular space. Since X admits C^1 triangulations, the ordinary homology of X can be computed with the complex of C^1-chains. So if we combine this with Stoke's theorem we get the *De Rham evaluation map*:

$$DR : H^p(\Gamma(X,\Omega_X^\cdot)) \to H^p(X,\mathbb{C}).$$

If X happens to be a Stein space, there is also a map going in the other direction. By general nonsense, there is a (first) spectral sequence of hypercohomology $E_1^{pq} = H^q(X,\Omega_X^p)$ converging to $\mathbb{H}^{p+q}(X,\Omega_X^\cdot)$. Since X is Stein, $E_1^{pq} = 0$ for $q > 0$ and so $\mathbb{H}^p(X,\Omega_X^\cdot) \cong H^p(\Gamma(X,\Omega_X^\cdot))$. Now think of $\mathbb{C}_X$ as a complex of sheaves on X (with trivial sheaves on the spots with a nonzero index). Then the inclusion $\mathbb{C}_X \to \Omega_X^\cdot$ is a map of complexes of sheaves and so there is and induced map on hypercohomology:

$$\alpha : H^p(X,\mathbb{C}) \cong \mathbb{H}^p(X,\mathbb{C}_X) \to \mathbb{H}^p(X,\Omega_X^\cdot) \cong H^p(\Gamma(X,\Omega_X^\cdot)).$$

We claim that α is a section of DR. To see this, we need the sheaf C_X^p on X associated to $X \supset U \to \{C^1\text{-p-cochains on } U \text{ with values in } \mathbb{C}\}$. These sheaves form in an obvious way a complex $C_X^\cdot$ on X. It is clear that integration yields a sheafified De Rham evaluation homomorphism $\Omega_X^\cdot \to C_X^\cdot$ of complexes which induces De Rham evaluation on hypercohomology. Since the composite morphism $\mathbb{C}_X \to \Omega_X^\cdot \to C_X^\cdot$ is an acyclic resolution for $\mathbb{C}_X$, it follows that the induced map on hypercohomology is the identity. So $DR\circ\alpha = \mathbf{1}$ as asserted.

In the (relative) case of a morphism $f : X \to S$ of analytic spaces, we get similarly for any $s \in S$ a De Rham evaluation map

$$DR_s : H^p(f_*\Omega^\cdot_f)_s \to H^p(f^{-1}(s),\mathbb{C})$$

Unless f is topologically locally trivial, these will not, in general, fit together to define a sheaf homomorphism from $H^p(f_*\Omega^\cdot_f)$ to $(R^pf_*\mathbb{C}_X) \otimes_{\mathbb{C}} O_S$.

(8.5) *Proposition*. Let $f : X \to S$ be a good Stein representative of a deformation $f : (X,x) \to (\mathbb{C}^k,0)$ of an isolated singularity. Then there is a natural exact sequence of O_S-modules

$$..\to R^pf_*H^0(\Omega^\cdot_f) \overset{\alpha^p}{\to} H^p(f_*\Omega^\cdot_f) \overset{\beta^p}{\to} f_*H^p(\Omega^\cdot_f) \to R^{p+1}f_*H^0(\Omega^\cdot_f) \to ..$$

which begins with $0 \to R^1f_*H^0(\Omega^\cdot_f) \to ..$. The composite of the natural map $i^p : R^pf_*\mathbb{C}_X \to R^pf_*H^0(\Omega^\cdot_f)$ with α^p is in any $s \in S$ a section of the De Rham evaluation map $DR_s : H^p(f_*\Omega^\cdot_f)_s \to H^p(X_s,\mathbb{C})$. Finally, if $X' \subset X$ is such that $f|X'$ is also a good Stein representative then the restriction homomorphism between the corresponding exact sequences is an isomorphism.

(8.6) *Corollary*. In the situation of (8.5), the O_S-homomorphism

$$(R^pf_*\mathbb{C}_X) \otimes_{\mathbb{C}} O_S \to H^p(f_*\Omega^\cdot_f)$$

induced by $\alpha^p \circ i^p$ is over $S-D_f$ an isomorphism. In $0 \in S$, the map

$$\beta^p : H^p(f_*\Omega^\cdot_f)_0 \to f_*H^p(\Omega^\cdot_f)_0 = H^p(\Omega^\cdot_{f,x})$$

is an isomorphism for $p > 0$.

Proof. As $H^p(\Omega^\cdot_f)$, $p > 0$, is supported by C_f, $f_*H^p(\Omega^\cdot_f)$ vanishes outside D_f. Moreover, $H^0(\Omega^\cdot_f) = f^{-1}O_S$ over $S-D_f$. So α^p induces an isomorphism $R^pf_*f^{-1}O_S \to H^p(f_*\Omega^\cdot_f)$ over $S-D_f$. Since $R^pf_*f^{-1}O_S \cong (R^pf_*\mathbb{C}_X) \otimes_{\mathbb{C}} O_S$ by lemma (8.1), the first part follows.

The last sentence of (8.5) implies that the stalk of $R^pf_*H^0(\Omega^\cdot_f)$ in $0 \in S$ is the inductive limit of the stalks $(R^pf_*H^0(\Omega^\cdot_f|X'))_0$, where X' runs over neighbourhoods of x in X such that $f : X' \to f(X')$ is a good Stein representative. Since such X' form a neighbourhood basis of x, it follows that $(R^p f_* \mathcal{H}^0(\Omega_f^\cdot))_0 = \lim_{U \ni 0} H^p(U, \mathcal{H}^0(\Omega_f^\cdot)) = 0$ for $p > 0$. So β^p is in 0 an isomorphism for $p > 0$.

Proof of (8.5). We have two (standard) spectral sequences

$$'E_2^{pq} = \mathcal{H}^p(R^q f_* \Omega_f^\cdot),\ ''E_2^{pq} = R^p f_* \mathcal{H}^q(\Omega_f^\cdot)$$

both converging to $R^{p+q} f_*(\Omega_f^\cdot)$. Since f is Stein, $R^q f_* \Omega_f^p = 0$ for $q > 0$ and so the first sequence degenerates. Hence $''E_2^{pq}$ converges to $\mathcal{H}^p(f_* \Omega_f^\cdot)$. For $q > 0$, $\mathcal{H}^q(\Omega_f^\cdot)$ is concentrated on the critical locus C_f of f. As $f|C_f$ is finite, hence Stein, it follows that $''E_2^{pq} = 0$ for $p,q > 0$. Then $''E_\infty^{pq}$ is easily computed: it is trivial if $p,q > 0$, equal to $''E_2^{pq}$ for $(p,q) = (0,0),(1,0)$ and for $p \geq 2$ there is an exact sequence

$$0 \to {''E_\infty^{0,p-1}} \to {''E_2^{0,p-1}} \xrightarrow{d_p} {''E_2^{p,0}} \to {''E_\infty^{p,0}} \to 0.$$

These combine to give the exact sequence of the proposition. The discussion preceding this proposition implies that for a fixed $s \in S$, the diagram

$$\begin{array}{ccc} (R^p f_* \mathcal{H}^0(\Omega_f^\cdot))_s & \xrightarrow{\alpha^p} & \mathcal{H}^p(f_* \Omega_f^\cdot)_s \\ \uparrow i^p & & \downarrow \\ H^p(X_s, \mathbb{C}) \cong (R^p f_* \mathbb{C}_X)_s & \overset{DR}{\rightleftarrows} & H^p(\Gamma(X_s, \Omega_{X_s}^\cdot)) \end{array}$$

commutes. It follows that $\alpha^p \circ i^p$ defines a section of the De Rham evaluation map. For the last part of the proposition we observe that it suffices to show that the maps $R^p f_* \mathcal{H}^0(\Omega_f^\cdot) \to R^p f_* \mathcal{H}^0(\Omega_f^\cdot|X')$ are isomorphisms, (because the maps $f_* \mathcal{H}^p(\Omega_f^\cdot) \to f_* \mathcal{H}^p(\Omega_f^\cdot|X')$ are clearly isomorphisms for $p > 0$ and the result then follows from the five lemma). Without loss of generality we may suppose that $\overline{X'} \subset X$; Let $N \subset X - C_f$ be open such that $X' \cup N = X$ and $f : X-N \to S$ is also a good representative. Then $(N, N \cap X')$ is a topologically trivial fibre bundle over S with fibre homeomorphic to $(\partial \overline{X}_0' \times (0,1), \partial \overline{X}_0' \times (0,\frac{1}{2}))$. It then follows from lemma (8.2) that the natural map $R^p f_*(f^{-1}\mathcal{O}_S | N \cap X) \to R^p f_*(f^{-1}\mathcal{O}_S | N)$ is an isomorphism. Feeding this into the Mayer-Vietoris sequence of the covering $\{N, X'\}$ of X with respect to the sheaf $H^0(\Omega_f^{\cdot})$ then gives that $R^p f_* H^0(\Omega_f^{\cdot}) \to R^p f_* H^0(\Omega_f^{\cdot} | X')$ is an isomorphism.

We now come to one of the more fundamental results of the theory.

(8.7) *Coherence theorem*. For a good Stein representative $f : X \to S$ of a deformation of an isolated singularity, the sheaves $H^p(f_* \Omega_f^{\cdot})$ are coherent $\mathcal{O}_S$-modules.

If f were proper, then it would follow from the famous Grauert coherence theorem that the sheaves $f_* \Omega_f^p$ would be coherent and hence so would be the $H^p(f_* \Omega_f^{\cdot})$. This fact was exploited by Brieskorn (1970a) for his proof of (8.7) in case f represents a germ $(\mathbb{C}^{n+1}, 0) \to (\mathbb{C}, 0)$ (in that case f can be taken to be the restriction of a projective morphism). Hamm subsequently generalized this using the Hironaka resolution theorem. There is also a proof which does not require that f extends to a proper morphism. As Buchweitz & Greuel (1980) observed, the coherence theorem can be derived from one of the main results (Thm. 3.7) in Kiehl & Verdier (1971). This "theorem of Schwartz type" is for Kiehl and Verdier the functional-analytic key to their proof of the Grauert coherence theorem. As its very formulation requires an extensive discussion of nuclear spaces and related concepts we do not quote it here.

Suffice it to say that the only non trivial input is furnished by prop. (8.5): if $X' \subset X$ is such that $\overline{X}'$ is proper over S and $f : X' \to S$ is a good Stein representative then the restriction map $H^p(f_*\Omega^{\cdot}_f) \to H^p(f_*(\Omega^{\cdot}_f|X'))$ is an isomorphism. For the rest of the argument we refer to §2 of Buchweitz & Greuel. (They assume the base S to be one-dimensional, but this hypothesis is not used in that section.) It is also possible to derive the coherence theorem using ingredients of the proof Forster & Knorr (1971) give of Grauert's theorem. This is probably simpler and certainly more direct than the other proof.

We warn the reader that the other terms of the exact sequence of (8.5) are in general not coherent $\mathcal{O}_S$-modules.

We mention an interesting consequence of the coherence theorem.

(8.8) *Corollary*. Let $f : (X,x) \to (\mathbb{C}^k,0)$ be a deformation of an isolated singularity of dim $n > 0$. Then for $p > 0$, $H^p(\Omega^{\cdot}_{f,x})$ is a noetherian $\mathcal{O}_{\mathbb{C}^k,0}$-module, and if f is a smoothing, the rank of $H^p(\Omega^{\cdot}_{f,x})$ is just the pth Betti number of a Milnor fibre.

Proof. By (8.6) $H^p(\Omega^{\cdot}_{f,x})$ is for $p \geq 1$ just the stalk of $H^p(f_*\Omega^{\cdot}_f)$ in $0 \in S$, where $f : X \to S$ is some good Stein representative. As $H^p(f_*\Omega^{\cdot}_f)$ is coherent, $H^p(\Omega^{\cdot}_{f,x})$ is noetherian. If r denotes its rank, then there is a neighbourhood V of 0 in S and an $\mathcal{O}_V$-homomorphism $\mathcal{O}_V^r \to H^p(f_*\Omega^{\cdot}_f)|V$ whose kernel and cokernel are supported by a proper subvariety of V. If we choose $s \in V$ outside D_f and this subvariety, then by (8.6), $(R^pf_*\mathbb{C}_X)_s \otimes_{\mathbb{C}} \mathcal{O}_{S,s}$ is a free $\mathcal{O}_{S,s}$-module of rank r. As $(R^pf_*\mathbb{C}_X)_s = H^p(X_s,\mathbb{C})$, r must be the dimension of $H^p(X_s,\mathbb{C})$.

8.B *The Gauss-Manin connection*

(8.9) Let M be an analytic manifold and let V be a holomorphic vector bundle over M. We denote the sheaf of holomorphic section of V by $\mathcal{O}(V)$. A *connection* on V tells you how to differentiate local sections of V in a tangent direction on M. Formally, it is a $\mathbb{C}$-linear map $\nabla : \mathcal{O}(V) \to \Omega_M \otimes_{\mathcal{O}_M} \mathcal{O}(V)$ satisfying the Leibniz rule: if $\phi \in \mathcal{O}_M$ and $\sigma \in \mathcal{O}(V)$, then $\nabla(\phi\sigma) = \phi\nabla\sigma + d\phi\otimes\sigma$. A derivation $\xi \in \Theta_{M,p}$ then induces a derivation $\nabla_\xi : \mathcal{O}(V)_p \to \mathcal{O}(V)_p$, $\sigma \to \langle\xi,\nabla\sigma\rangle$, called the *covariant derivative* with respect to ξ.

A local section $\sigma \in \mathcal{O}(V)_p$ is called *horizontal* if $\nabla\sigma = 0$. In terms of a local coordinate system $z_1,\ldots,z_m$ for (M,p) and a local frame $\sigma_1,\ldots,\sigma_r$ of V at p, the condition that $\sigma = \phi_1\sigma_1 +\ldots+ \phi_r\sigma_r$ $(\phi_\rho \in \mathcal{O}_{M,p})$ be horizontal is that

$$0 = \nabla_{\partial/\partial z_\mu}(\sigma) = \Sigma_{\rho=1}^{r} \frac{\partial\phi_\rho}{\partial z_\mu}\sigma_\rho + \phi_\rho\nabla_{\partial/\partial z_\mu}(\sigma_\rho) \quad (\mu=1,\ldots,m)$$

Writing $\nabla_{\partial/\partial z_\mu}(\sigma_\rho)$ as an $\mathcal{O}_{M,p}$-linear combination of $\sigma_1,\ldots,\sigma_r$, we find that mr (first order) differential equations must be satisfied, while we have only r degrees of freedom. So local horizontal sections will in general not exist unless m = 1. We say that ∇ is *integrable* if the horizontal sections generate $\mathcal{O}(V)$. It is not difficult to show that this is equivalent to its infinitesimal form:

$$\nabla_{[\xi_1,\xi_2]} = \nabla_{\xi_1}\nabla_{\xi_2} - \nabla_{\xi_2}\nabla_{\xi_1} \quad \text{for all } \xi_1,\xi_2 \in \Theta_M$$

which can also be phrased as: the composite $R(\nabla) : \mathcal{O}(V) \xrightarrow{\nabla} \Omega^1_M \otimes \mathcal{O}(V) \xrightarrow{1\otimes\nabla} \Omega^1_M \otimes \Omega^1_M \otimes \mathcal{O}(V) \xrightarrow{\wedge\otimes 1} \Omega^2_M \otimes \mathcal{O}(V)$ is trivial ($R(\nabla)$ is the so-called *curvature homomorphism* of ∇). The uniqueness theorem for ODE's implies that there is at most one horizontal $\sigma \in \mathcal{O}(V)_p$

with given $\sigma(p)$. So if ∇ is integrable, the horizontal sections provide *natural* local trivializations of V and they define a local system $\mathbb{V}$ on M with $\mathcal{O}_M \otimes_{\mathbb{C}} \mathbb{V} \cong \mathcal{O}(V)$. Conversely, any local system $\mathbb{V}$ on M determines a holomorphic vector bundle V with integrable connection ∇: we have $\mathcal{O}(V) = \mathcal{O}_M \otimes_{\mathbb{C}} \mathbb{V}$ and if $\sigma_1,\ldots,\sigma_r$ is a basis for $\mathbb{V}_p$, then we define ∇ by $\nabla(\phi_1\sigma_1+\ldots+\phi_r\sigma_r) = d\phi_1 \otimes \sigma_1 +\ldots+ d\phi_r \otimes \sigma_r$.

Our main example is furnished by a good Stein representative $f : X \to S$ of a deformation of an isolated singularity. Over S-D, f is C^∞-locally trivial and so $R^pf_*\mathbb{C}_X$ is a local system over S-D. Therefore, $R^pf_*\mathbb{C}_X \otimes_{\mathbb{C}} \mathcal{O}_{S-D}$ is in a natural way endowed with an integrable connection. As $\mathcal{H}^p(f_*\Omega^\cdot_f)$ and $R^pf_*\mathbb{C}_X \otimes_{\mathbb{C}} \mathcal{O}_S$ are naturally isomorphic over S-D, this integrable connection corresponds to one on $\mathcal{H}^p(f_*\Omega^\cdot_f)|S\text{-}D$ (provisionally called the *topological connection*). We describe this connection in terms of the relative p-forms and use this to extend the connection (meromorphically) to all of $\mathcal{H}^p(f_*\Omega^\cdot_f)$.

First recall the definition of the Lie derivative. If (M,p) is an analytic germ and $\xi \in \theta_{M,p}$, then "*contraction with* ξ": $\iota_\xi : \Omega^\cdot_{M,p} \to \Omega^{\cdot-1}_{M,p}$ is the unique square zero $\mathcal{O}_{M,p}$-linear anti-derivation of degree one which on $\Omega^1_{M,p}$ is just given by $\xi : \Omega^1_{M,p} \to \mathcal{O}_{M,p}$. The *Lie-derivative with respect to* ξ is defined by $L_\xi = \iota_\xi \circ d + d \circ \iota_\xi : \Omega^\cdot_{M,p} \to \Omega^\cdot_{M,p}$. One easily checks that L_ξ is a $\mathbb{C}$-derivation of degree 0 which on $\mathcal{O}_{M,p}$ is just the derivation defined by ξ. Moreover, $\xi \in \theta_{M,p} \mapsto L_\xi \in \mathrm{End}(\Omega^\cdot_{M,p})$ is a Lie-homomorphism ($L_{[\xi_1,\xi_2]} = L_{\xi_1} \circ L_{\xi_2} - L_{\xi_2} \circ L_{\xi_1}$) and L_ξ commutes with d. It is clear that L_ξ takes closed r-forms to exact r-forms and thus acts trivially on $H^r(\Omega^\cdot_{M,p})$. The geometric meaning of L_ξ can be understood as follows: if $h_t : (M,p) \to M$ (t in a neighbourhood of 0 in $\mathbb{C}$) is the flow generated by ξ, and $\omega \in \Omega^r_{M,p}$, then

$$L_\xi(\omega) = \lim_{t\to 0} \frac{1}{t}(h_t^*(\omega)-\omega).$$

Returning to our Stein representative, let $s \in S$ and $\eta \in \theta_{S,s}$ be such that η can be lifted to a $\xi \in (f_*\theta_X)_s$. If $s \in S-D$ this does not impose a condition on η, for the sequence of coherent $\mathcal{O}_X$-modules

$$0 \to \theta_f \to \theta_X \xrightarrow{\partial f} \mathcal{O}_X \otimes_{\mathcal{O}_S} \theta_S \to 0$$

is then exact over a neighbourhood of s and hence its f-direct image is exact in s (for f Stein implies $R^1 f_*\theta_f = 0$). The same type of argument shows that we have an exact sequence

$$0 \to \Sigma_{\kappa=1}^{k} df_\kappa \wedge f_*\Omega_X^{\cdot -1} \xrightarrow{\pi} f_*\Omega_X^{\cdot} \to f_*\Omega_f^{\cdot} \to 0.$$

If $\omega \in f_*\Omega_X^{p-1}$, then

$$L_\xi(df_\kappa \wedge \omega) = L_\xi(df_\kappa) \wedge \omega + df_\kappa \wedge L_\xi(\omega)$$

Since $L_\xi(df_\kappa) = d(\xi(f_\kappa)) = f^*d(\eta(t_\kappa))$, it follows that L_ξ leaves the image of π invariant and thus induces a derivation of $f_*\Omega_f^{\cdot}$ (which we continue to denote by L_ξ). Since L_ξ commutes with d_X, L_ξ also commutes with d_f, but L_ξ need not send a d_f-closed form to a d_f-exact form. However if ξ is vertical (i.e. $\eta = 0$), then this is the case: if $\omega \in (f_*\Omega_X^p)_s$ represents a d_f-closed form in $(f_*\Omega_f^p)_s$, then $d\omega = \Sigma_{\kappa=1}^{k} df_\kappa \wedge \omega_\kappa$ for certain $\omega_\kappa \in (f_*\Omega_X^p)_s$ and so

$$\begin{aligned} L_\xi(\omega) &= \iota_\xi d(\omega) + d\iota_\xi(\omega) \\ &= \iota_\xi(\Sigma_{\kappa=1}^{k} df_\kappa \wedge \omega_\kappa) \bmod df_*\Omega_X^{p-1} \\ &= \Sigma_{\kappa=1}^{k} -df_\kappa \wedge \iota_\xi(\omega_\kappa) \bmod df_*\Omega_X^{p-1}. \end{aligned}$$

As two lifts of η differ by a vertical vector field, it follows that the action of L_ξ on $H^p(f_*\Omega_f^{\cdot})_s$ only depends on η. We denote this action by ∇_η. Notice that ∇_η has the formal properties of an integrable connection: it is a $\mathbb{C}$-derivation of $H^{\cdot}(f_*\Omega_f^{\cdot})$ of degree 0 satisfying

$$\nabla_\eta([\omega_1]\wedge[\omega_2]) = \nabla_\eta([\omega_1])\wedge[\omega_2]+[\omega_1]\wedge\nabla_\eta([\omega_2])$$
$$\nabla_\eta(\phi) = \eta(\phi) \qquad \text{if } \phi \in \mathcal{O}_{S,s} (\subset \mathcal{H}^0(f_*\Omega_f^\cdot))$$
$$\nabla_{[\eta_1,\eta_2]} = \nabla_{\eta_1}\nabla_{\eta_2} - \nabla_{\eta_2}\nabla_{\eta_1}$$

(this is immediate from the corresponding properties of L_ξ). Although $\eta \to \nabla_\eta$ is not a connection in the sense of our definition ($\eta \in \theta_{S,s}$ is not arbitrary and $\mathcal{H}^p(f_*\Omega_f^\cdot)$ needn't be a locally free $\mathcal{O}_S$-module), it is so over S-D. If $s \in S-D$ and $\omega \in (f_*\Omega_f^p)_s$ is d_f-closed, then write $d\omega = df_1 \wedge \omega_1 + \ldots + df_k \wedge \omega_k$ as before. If $\xi_\kappa \in f_*\theta_X$ is a lift of $\partial/\partial t_\kappa$, then

$$L_{\xi_\kappa}(\omega) \equiv \iota_{\xi_\kappa} d(\omega) \bmod d(f_*\Omega_X^{p-1})$$
$$\equiv \omega_\kappa \bmod(\Sigma_{\lambda=1}^k df_\lambda \wedge f_*\Omega_X^{p-1} + df_*\Omega_X^{p-1})$$

and so ∇ is then given by

$$\text{(8.9a)} \qquad \nabla([\omega]) = dt_1 \otimes [\omega_1] + \ldots + dt_k \otimes [\omega_k].$$

We claim that ∇ coincides over S-D with the topological connection introduced earlier. To see this, let V be a convex neighbourhood of s in S-D such that $\omega, \omega_1, \ldots, \omega_k$ converge on X_V and f admits a C^∞-trivialization $h : X_V \to X_s$ over V. Given a p-cycle Z on X_s and $s' \in V$, let $Z_{s'}$ be the p-cycle on $X_{s'}$ defined by $(h|X_{s'})^{-1}Z_s$. Similarly, if $s_0, s_1 \in V$, let $Z_{[s_0,s_1]}$ denote the (p+1)-chain on X_V defined as the pre-image of $[s_0,s_1]\times Z$ under $(f,h) : X_V \to V\times X_s$. Then $\partial Z_{[s_0,s_1]} = Z_{s_1} - Z_{s_0}$, so that according to Stokes

$$\int_{Z_{s_1}} \omega - \int_{Z_{s_0}} \omega = \int_{Z_{[s_0,s_1]}} d\omega = \int_{Z_{[s_0,s_1]}} \Sigma_\kappa df_\kappa \wedge \omega_\kappa$$
$$= \Sigma_\kappa \int_0^1 df_\kappa \int_{Z_{(1-t)s_0+ts_1}} \omega_\kappa$$

It follows that $s' \in V \to \int_{Z_{s'}} \omega$ is constant if and only if $\int_{Z_{s'}} \omega_\kappa = 0$ for $s' \in V$, $\kappa = 1,\dots,k$. By letting Z run over all homology classes we see that being horizontal for the topological connection is the same as being horizontal for the Gauss-Manin connection. Hence both connections coincide on S-D.

So far we defined ∇_η only for liftable η. In order to define ∇_η for all η we introduce the submodule $\underset{\sim}{\theta}_f \subset \theta_S$ of *liftable vector fields*; in other words, $\underset{\sim}{\theta}_f$ is the kernel of the natural map

$$\theta_S \to f_*\mathcal{O}_X \otimes_{\mathcal{O}_S} \theta_S/\partial f(f_*\theta_X)$$

We check that $\underset{\sim}{\theta}_f$ is coherent. As f is Stein, the right hand side may be identified with the direct image of $\mathcal{O}_X \otimes_{\mathcal{O}_S} \theta_S/\partial f(\theta_X)$ under f. The latter is what we denoted by T_f in the case of a nonsingular X; it is a coherent $\mathcal{O}_X$-module supported by C_f. Since $f|C_f$ is finite, $f_*(\mathcal{O}_X \otimes_{\mathcal{O}_X} \theta_S)/\partial f(\theta_X)$ is a coherent $\mathcal{O}_S$-module and hence so is $\underset{\sim}{\theta}_f$. We now assume that D_{red} is a hypersurface in S which can be defined by a single equation $\delta \in \Gamma(S,\mathcal{O}_S)$. The De Rham cohomology sheaf $\mathcal{H}^p(f_*\Omega_f^\cdot)$ is locally free outside D and so $\mathcal{H}^p(f_*\Omega_f^\cdot)[\delta^{-1}]$ is a locally free $\mathcal{O}_S[\delta^{-1}]$-module ($[\delta^{-1}]$ is short for $\otimes_{\mathcal{O}_S} \mathcal{O}_S[\delta^{-1}]$). Since $\theta_S/\underset{\sim}{\theta}_f$ is supported by D, there exists for each $s \in S$ an $\ell \in \mathbb{N}$ so that $\delta^\ell \theta_{S,s} \subset \underset{\sim}{\theta}_{f,s}$. So if $\eta \in \theta_{S,s}$, then $\nabla_{\delta^\ell \eta}$ is defined and the Leibniz rule tells us what ∇_η should be:

$$\nabla_\eta([\omega]) = \delta^{-\ell} \nabla_{\delta^\ell \eta}([\omega]) \in \mathcal{H}^p(f_*\Omega_f^\cdot)[\delta^{-1}].$$

It is easy to check that $\nabla_\eta([\omega])$ is independent of the choice of δ and ℓ. So the Gauss-Manin connection now appears as a connection on $\mathcal{H}^p(f_*\Omega_f^\cdot)[\delta^{-1}]$:

$$\nabla : \mathcal{H}^p(f_*\Omega_f^\cdot)[\delta^{-1}] \to \Omega_S^1 \otimes_{\mathcal{O}_S} \mathcal{H}^p(f_*\Omega_f^\cdot)[\delta^{-1}].$$

Alternatively, if $\omega \in (f_*\Omega_f^p)_s$ is d_f-closed and $d\omega = \Sigma_\kappa df_\kappa \wedge \omega_\kappa$ then $\delta^\ell \omega_\kappa$ is d_f-closed for some ℓ ($\kappa=1,\dots,k$) and so formula (8.9a) continues to hold. It is clear that this formula can also be used to define ∇ on $f_*\mathcal{H}^p(\Omega_f^\cdot)$. In particular, we have a connection on $H^p(\Omega_{f,x}^\cdot)[\delta^{-1}]$.

*(8.10) An important property of ∇ is that it is regular-singular along D. Although we shall not make any use of this fact, we don't want to leave it unmentioned as it is one of the points of departure of the recent applications of the theory of $\mathcal{D}$-modules to singularity theory. In order to define this property, we start out with an analytic manifold M, a hypersurface $H \subset M$ and a locally free $\mathcal{O}_M[\mathcal{I}_H^{-1}]$-module $\mathcal{F}$ of finite constant rank r. A section σ of $\mathcal{F}$ over some open $A \subset M-H$ is said to have *tempered growth* if for all $p \in \overline{A}$ there exists a basis $\sigma_1,\dots,\sigma_r$ of $\mathcal{F}$ over some neighbourhood U of p (in M, of course) such that the coefficients of σ in $\sigma_1,\dots,\sigma_r$ are bounded functions on $A \cap U$. If $\nabla : \mathcal{F} \to \Omega_M \otimes_{\mathcal{O}_M} \mathcal{F}$ is an integrable connection on $\mathcal{F}$ (the definition being the obvious one), then the following two properties are equivalent

(i) The horizontal sections of ∇ over open subsets of M-H have tempered growth.

(ii) Any stalk $\mathcal{F}_p$ contains a noetherian $\mathcal{O}_{M,p}$-submodule which generates $\mathcal{F}_p$ over $\mathcal{O}_{M,p}[\mathcal{I}_M^{-1}]$ and is invariant under $\theta_{M,p}\langle H\rangle$ (the action being given by $\eta \in \theta_{M,p}\langle H\rangle : \sigma \in \mathcal{F}_p \to \nabla_\eta \sigma \in \mathcal{F}_p$).

The implication (i) $\Rightarrow$ (ii) follows from Deligne (1970), Thm. II. 4.1 and Van den Essen (1979) IV. 2.3, for the converse we refer also to Van den Essen (1979) II. 5.1. (I am indebted to Van den Essen for these references.) Notice that property (i) is clearly invariant under base change (in a sense we leave to the reader to make precise), whereas for

(ii) this is by no means obvious. If either one of these properties is satisfied, we say that ∇ has a *regular singularity along* H.

(8.11) *Theorem*. The Gauss-Manin connection has a regular singularity along D.

Proof. We follow the line of argument used by Malgrange (1974b) in the hypersurface case. It follows from the stratification theory of Thom (1969) (see also Gibson et al. (1976), Ch. I) and the triangulation theorems of Łojasiewicz (see Hironaka, 1974) that $\overline{X}$ and S admit arbitrarily fine C^1-triangulations $\tilde{T}$ and T such that f becomes a simplicial map and D is a subcomplex of S. Let σ be any ℓ-simplex of T with $|\sigma| \not\subset D$ and suppose T fine enough in order that there exist $\omega_1,\ldots,\omega_r \in \Gamma(|\sigma|, f_*\Omega_f^p)$ mapping to a basis of $H^p(f_*\Omega_f^{\cdot})[\delta^{-1}]$ over $|\sigma|$. If $\tilde{\sigma}$ is any $(p+\ell)$-simplex of $\tilde{T}$ lying over σ then the expressions

$$(*) \qquad \int_{|\tilde{\sigma}|\cap\overline{X}_s} \omega_i$$

are well-defined for all $s \in |\sigma|$ and define continuous functions on $|\sigma|$.

Let V be a compact neighbourhood of $|\sigma|$ in S over which $\omega_1,\ldots,\omega_r$ are still defined. For any $s \in V-(V\cap D)$ we choose an integral basis $c_1(s),\ldots,c_r(s)$ of $H_p(\overline{X}_s;\mathbb{Z})$/torsion and form

$$d(s) := \det{}^2\ (\langle c_i(s), [\omega_j(s)]\rangle).$$

Clearly, d(s) is independent of the basis chosen and d is a holomorphic nowhere zero function on $V-(V\cap D)$. If we restrict d to the intersection of $V-(V\cap D)$ with a simplex of T, then we get a squared determinant whose entries are finite sums of expressions of the type (*). This implies that d is bounded on V. Hence d extends holomorphically to V and there exists an $\ell \in \mathbb{N}$ such that $d^{-1}\delta^{\ell}$ is holomorphic on V.

If $\zeta_1,\ldots,\zeta_r$ is a basis of horizontal sections of $H^p(f_*\Omega^{\cdot}_f)$ over $|\sigma|-(|\sigma|\cap D)$, then write

$$\zeta_i = \Sigma_j\ \phi_{ij}[\omega_j],\ \phi_{ij} \in \Gamma(|\sigma|-(|\sigma|\cap D),\mathcal{O}_S).$$

We will show that the functions $d.\phi_{ij}$ extend continuously over $|\sigma|$; this certainly implies that the ζ_i have tempered growth (use the basis $\{d^{-1}[\omega_j]\}$). If $e_1(s),\ldots,e_r(s)$ is the basis of $H_p(\overline{X}_s;\mathbf{C})$ dual to $\zeta_1(s),\ldots,\zeta_r(s)$, then it follows from the continuity of the expressions of the form (*) that the functions $s \in (|\sigma|-|\sigma|\cap D) \to \langle e_i(s),[\omega_j(s)]\rangle$ extend continuously over $|\sigma|$. Now consider the equality of matrices

$$\delta_{ij} = \langle e_i(s),\zeta_j(s)\rangle = \Sigma_h\langle e_i(s),\omega_h(s)\rangle\phi_{jh}(s).$$

Since $\det^2\langle e_i(s),\omega_h(s)\rangle$ is a scalar multiple of d and $\langle e_i(s),\omega_h(s)\rangle$ is continuous on $|\sigma|$, it follows from Cramer's rule that $d.\phi_{jh}$ is continuous on $|\sigma|$.

(8.12) For complete intersections there is an alternative proof of theorem (8.11) which uses the equivalence between the two defining properties of regularity: since the property of being regular-singular is invariant under base change it suffices to consider the case of a Thom-transversal deformation. According to (6.14) we then have $\theta_S\langle D\rangle = \underset{\sim}{\theta}_f$ (at least on the germ level but the fact that f is Stein implies that this is also globally true). Since $\underset{\sim}{\theta}_f$ leaves $H^p(f_*\Omega^{\cdot}_f)$ invariant, property (ii) is satisfied in a rather trivial way.

In the case of a germ $f : (\mathbf{C}^{n+1},x) \to (\mathbf{C},0)$, the regularity theorem was proved by Brieskorn (1970) using a corresponding global result due to Griffiths. He pointed out that this leads to an algorithm for computing the homological monodromy of f (up to conjugacy). He also showed that the formation of $(H^n(\Omega^{\cdot}_{f,p}),\nabla)$ is algebraic, i.e. behaves well under field automorphisms of $\mathbb{C}$. Combining this in a clever way with the Gelfond-Schneider theorem (which says: if α and $e^{2\pi i\alpha}$ are algebraic, then $\alpha \in \mathbb{Q}$) gave the quasi-unipotence of the monodromy of f. Generalizations and alternative proofs of the regular singularity were obtained by Deligne (1970), Saito (1973) and Greuel (1975).

8.C *The complete intersection case*

(8.13) We begin with recalling some basic facts concerning local cohomology. For more information we refer to the lecture notes of Grothendieck (1967) and Siu-Trautmann (1971).

Let X be a space and $Z \subset X$ a locally closed subspace (which means that Z is closed in some neighbourhood of Z in X). For any abelian sheaf F over X and any neighbourhood U of Z in X in which Z is closed, the subgroup of $s \in H^0(U,F)$ with support contained in Z does not depend on U. We denote this subgroup by $H^0_Z(F)$. This defines a left-exact functor from the category of abelian sheaves over X to the category of abelian groups. As the former has enough injectives, we have the right derived functors of $H^0_Z(\ )$, denoted $H^q_Z(\ )$, $q = 0,1,2,\ldots$; the group $H^q_Z(F)$ is called the qth *cohomology group of* F *with supports in* Z. For reasonable spaces X (e.g. X paracompact and locally contractible: the kind of spaces we will always be dealing with) $H^q_Z(F)$ may be identified with the relative (Čech) cohomology group $H^q(X,X-Z;F)$. This concept can be sheafified: $\mathcal{H}^q_Z(\ )$ is the sheaf associated to the presheaf: U open in $X \to H^q_{Z\cap U}(F|U)$. Alternatively, $\mathcal{H}^q_Z(\ )$ may be defined as the qth right derived functor of the sheafified $H_Z^0(\ )$. It is not hard to show that there is an exact sequence

(8.13a) $$0 \to \mathcal{H}_Z^0(F) \to F \to j_*j^{-1}F \to \mathcal{H}_Z^1(F) \to 0$$

where $j : X-Z \subset X$ and that for $q \geq 2$

(8.13b) $$\mathcal{H}_Z^q(F)_x = \varinjlim\{H^{q-1}(U-(U\cap Z);F) : x \in U \text{ open}\}.$$

Clearly, $\mathcal{H}_Z^q(F)$ has its support contained in $\overline{Z}$. By general nonsense there is a spectral sequence

$$E_2^{pq} = H^p(X,\mathcal{H}_Z^q(F)) \Rightarrow H_Z^{p+q}(F).$$

By the very definition of a right derived functor, an exact sequence $0 \to F \to G \to H \to 0$ of abelian sheaves on X gives rise to a long exact sequence of abelian groups

$$0 \to H_Z^0(F) \to H_Z^0(G) \to H_Z^0(H) \to H_Z^1(F) \to \dots$$

and likewise one of abelian sheaves. We shall often use the following simple observation:

(8.14) Let $0 \to F \to G \xrightarrow{\beta} H \to 0$ be a sequence of abelian sheaves on X, exact on X-Z. Suppose β surjective and $\mathcal{H}_Z^0(F), \mathcal{H}_Z^1(F), \mathcal{H}_Z^0(G)$ all trivial. Then the sequence is exact on all of X.

This assertion follows in a straightforward manner from the exactness of (8.13a).

The case that will mainly concern us is when X is an analytic space, $Z \subset X$ a (closed) analytic subspace and F a coherent sheaf on X. Then the sheaves $\mathcal{H}_Z^q(F)$ are in a natural way $\mathcal{O}_X$-modules. The stalk of $\mathcal{H}_Z^q(F)$

at $x \in X$ only depends on the stalk F_x; this justifies writing $H^q_Z(F_x)$ instead of $\mathcal{H}^q_Z(F)_x$. These cohomology groups are intimately related to the notion of depth: if $I \subset \mathcal{O}_{X,x}$ is an ideal defining Z at x, then the I-depth of F_x is precisely the first q for which $H^q_Z(F_x) \neq 0$. Usually, we are given X and F and choose $Z \subset X$ in such a way that F is locally free on X-Z. Then the sheaf $\mathcal{H}^q_Z(F)$ is a coherent $\mathcal{O}_X$-module if q < codim Z. If $\mathcal{O}_{X,x}$ is Cohen-Macaulay, then $\mathrm{codim}_x Z = I\text{-depth } \mathcal{O}_{X,x}$, so that $H^q_Z(\mathcal{O}_{X,x}) = 0$ for $q < \mathrm{codim}_x Z$. Hence the groups $H^q_Z(F_x)$, $q < \mathrm{codim}_x Z$ measure how non-trivial the vector bundle on X-Z defined by F is near x. For $q = \mathrm{codim}_x Z$, $x \in Z$, $H^q_Z(\mathcal{O}_{X,x})$ is not finitely generated.

After this general discussion of local cohomology we return to singularities. The following lemma is our point of departure.

(8.15) *Lemma*. Let X be an analytic space in an open $U \subset \mathbb{C}^N$ and let $F : U \to \mathbb{C}^K$ be an analytic map. Suppose we are further given an analytic subspace $Z \subset X$ defined by an ideal $J_Z \subset \mathcal{O}_X$ which contains the singular locus of X. Then for every $x \in X$, $H^q_Z(\Omega^p_F \otimes \mathcal{O}_{X,x}) = 0$ if $p \leq \mathrm{codim}_x(C_F \cap X, X)$ and $p+q < J_{Z,x}\text{-depth } \mathcal{O}_{X,x}$. Or equivalently: if $p \leq \mathrm{codim}_x(C_f \cap X, X)$, then

$$J_{Z,x}\text{-depth}(\Omega^p_F \otimes \mathcal{O}_{X,x}) \geq J_{Z,x}\text{-depth}(\mathcal{O}_{X,x}) - p.$$

Proof. We use induction on K. If K = 0, then $\Omega^p_F \otimes \mathcal{O}_X = \Omega^p_U \otimes \mathcal{O}_X$ is a free $\mathcal{O}_X$-module and so there is nothing to prove. If K > 0 we also do induction over p. Again, the induction starts trivially for p = 0, so let $0 < p < J_{Z,x}\text{-depth } \mathcal{O}_{X,x}$. Write $F' = (F_1, \ldots, F_{K-1}) : U \to \mathbb{C}^{K-1}$ and consider the sequence of $\mathcal{O}_U$-modules

$$S^p : 0 \to \Omega^{p-1}_F \xrightarrow{dF_K \wedge} \Omega^p_{F'} \to \Omega^p_F \to 0.$$

If F is a submersion in $y \in U$, then a simple local computation shows that $dF_K\wedge$ is injective in y and hence S^p_y is exact. So $S^p \otimes_{\mathcal{O}_U} \mathcal{O}_X$ is exact on $X-(C_F \cap X)$. Our induction hypotheses imply the vanishing of $H^0_Z(\Omega^{p-1}_F \otimes \mathcal{O}_{X,x})$, $H^1_Z(\Omega^{p-1}_F \otimes \mathcal{O}_{X,x})$ and $H^0_Z(\Omega^p_{F'} \otimes \mathcal{O}_{X,x})$. So by (8.13) the sequence $S^p \otimes \mathcal{O}_{X,x}$ is exact. The associated long exact sequence

$$\ldots \to H^q_Z(\Omega^p_{F'} \otimes \mathcal{O}_{X,x}) \to H^q_Z(\Omega^p_F \otimes \mathcal{O}_{X,x}) \to H^{q+1}_Z(\Omega^{p-1}_F \otimes \mathcal{O}_{X,x}) \to \ldots$$

and the induction hypotheses give that $H^q_Z(\Omega^p_F \otimes \mathcal{O}_{X,x}) = 0$ for $p+q < J_{Z,x}$-depth $\mathcal{O}_{X,x}$.

We now consider the case where X is a complete intersection defined by some of the components of F:

(8.16) *Corollary*. Suppose that in (8.15), X is a complete intersection in U (of codim $K-k$) defined by the last $K-k$ components $F_{k+1},\ldots,F_K$ of F. If we put $f = (F_1,\ldots,F_k) : X \to \mathbb{C}^k$, then

(i) For $p \le \operatorname{codim}(C_f,X)$ and $p+q < \operatorname{codim}(Z,X)$, $H^q_Z(\Omega^p_f) = 0$

(ii) For $p < \operatorname{codim}(C_f,X)$, the $\mathcal{O}_X$-homomorphisms

$$\Omega^p_f \to \Omega^{p+k}_X, \qquad \Omega^p_f \to \Omega^{p+K}_U \otimes_{\mathcal{O}_U} \mathcal{O}_X$$

induced by exterior multiplication with $df_1 \wedge\ldots\wedge df_k$ and $dF_1 \wedge\ldots\wedge dF_K$ respectively, are injective.

Proof. (i) Since X is a complete intersection, it is Cohen-Macaulay and so the $J_{Z,x}$-depth of $\mathcal{O}_{X,x}$ equals $\operatorname{codim}_x(Z,X)$ and similarly for C_f. As $C_F \cap X = C_f$ and $\Omega^p_F \otimes_{\mathcal{O}_U} \mathcal{O}_X \cong \Omega^p_f$, (i) follows from (8.15).

(ii) A simple local computation shows that the two homomorphisms are injective outside C_f. So their kernels are supported by C_f. These must be trivial since (i) implies that $H^0_{C_f}(\Omega^p_f) = 0$ for $p < \operatorname{codim}(C_f,X)$ (take $Z = C_f$).

(8.17) If in the above corollary, F defines an icis (X_0,x) of dim $n := N-K$, then f is a deformation of (X_0,x). We know that either $\mathrm{codim}_x(C_f,X) = n+1$ (i.e. f is a smoothing of (X_0,x)) or $\mathrm{codim}_x(C_f,X) = n$ (all fibres of f are singular). The corollary implies that $H^q_{C_f}(\Omega^p_{f,x}) = 0$ for $p+q \le n$ in case of smoothing and for $p+q < n$ otherwise. In particular, $H^q_{\{x\}}(\Omega^p_{X_0,x}) = 0$ for $p+q < n$, in other words depth $\Omega^p_{X_0,x} \ge n-p$.

(8.18) In the situation of (8.16), let $j : X-C_f \subset X$ denote the inclusion and let $n := N-K$ be the fibre dimension of F. There is a natural $\mathcal{O}_X$-homomorphism $\Omega^n_f \to j_*j^{-1}\Omega^n_f$. We will define a submodule ω_f of $j_*j^{-1}\Omega^n_f$ which contains the image of Ω^n_f and has the virtue of being isomorphic to $\mathcal{O}_X$ - a property that makes ω_f easier to work with than Ω^n_f. The homomorphism $dF_1 \wedge \ldots \wedge dF_K \wedge : \Omega^n_f \to \Omega^N_U \otimes \mathcal{O}_X$ is an isomorphism outside C_f: $j_*j^{-1}\Omega^n_f \cong j_*j^{-1}(\Omega^N_U \otimes \mathcal{O}_X)$. So if we let ω_f be the pre-image of $\Omega^N_U \otimes \mathcal{O}_X$ in $j_*j^{-1}\Omega^n_f$, then $\omega_f \cong \Omega^N_U \otimes \mathcal{O}_X \cong \mathcal{O}_X$. The sheaf ω_f plays an important rôle in Grothendieck's duality theory; it is therefore called the *dualizing sheaf* of f. It is not hard to check that ω_f only depends on f (in particular not on the extension F of f) and is compatible with base change. In the absolute case: $k = 0$, we often write ω_X instead of ω_f. Clearly, the image of Ω^n_f in $\Omega^N_U \otimes \mathcal{O}_X$ is just $\Omega^N_U \otimes_{\mathcal{O}_U} C_f$ and so we have an exact sequence

$$0 \to H^0_{C_f}(\Omega^n_f) \to \Omega^n_f \to \omega_f \to \omega_f \otimes \mathcal{O}_{C_f} \to 0. \tag{8.18a}$$

When C_f is of codim ≥ 2 in X, then $j_*j^{-1}(\Omega^N_U \otimes \mathcal{O}_X) = \Omega^N_U \otimes \mathcal{O}_X$ (for X is then normal) and hence ω_f may be identified with $j_*j^{-1}\Omega^n_f$. So in that case, it follows from (8.13a) that $H^1_{C_f}(\Omega^n_f) \cong \omega_f \otimes \mathcal{O}_{C_f}$.

In case $\dim_x C_f = k-1$, then $H^0_{C_f}(\Omega^n_{f,x}) = 0$ (by (8.17)) and hence (8.18a) reduces to a short exact sequence

$$0 \to \Omega^n_{f,x} \to \omega_{f,x} \to \omega_{f,x} \otimes \mathcal{O}_{C_f,x} \to 0$$
$$\qquad\qquad \wr\| \qquad\qquad \wr\|$$
$$\qquad\qquad \mathcal{O}_{X,x} \qquad \mathcal{O}_{C_f,x}$$

Following Greuel (1975) we can use this sequence to give an alternative proof of the fact that $\mathcal{O}_{C_f,x}$ is Cohen-Macaulay. By (8.16), $H^q_{\{x\}}(\Omega^n_{f,x}) = 0$ for $q < k$ and $H^q_{\{x\}}(\mathcal{O}_{X,x}) = 0$ for $q < n+k$. The long exact cohomology sequence associated to the above sequence then gives $H^q_{\{x\}}(\mathcal{O}_{C_f,x}) = 0$ for $q < k-1$ or equivalently, depth $\mathcal{O}_{C_f,x} \geq k-1$.

We take a closer look at the De Rham cohomology sheaves. We first consider the case of a single fibre.

(8.19) *Proposition*. Let (X,x) be an icis of dim $n \geq 1$. Then the complex

$$0 \to \mathbf{C} \to \mathcal{O}_{X,x} \xrightarrow{d} \Omega^1_{X,x} \to \dots \xrightarrow{d} \Omega^n_{X,x}$$

is exact, $H^0_{\{x\}}(d\Omega^{n-1}_X) = 0$ and we have a natural chain of injections

$$H^1_{\{x\}}(d\Omega^{n-2}_X) \hookrightarrow H^2_{\{x\}}(d\Omega^{n-3}_X) \hookrightarrow \dots \hookrightarrow H^{n-1}_{\{x\}}(d\mathcal{O}_X) \hookrightarrow H^n_{\{x\}}(\mathbf{C}_X).$$

Proof. We suppose that X is a good representative for (X,x). Since $\mathcal{O}_{X,x}$ is reduced,

$$S^0 : 0 \to \mathbf{C} \to \mathcal{O}_{X,x} \xrightarrow{d} d\mathcal{O}_{X,x} \to 0$$

is exact. Suppose inductively that we have shown that for some $m \leq n-1$

$$S^p : 0 \to d\Omega^{p-1}_{X,x} \to \Omega^p_{X,x} \to d\Omega^p_{X,x} \to 0$$

is exact for $p < m$. Since $H^q_{\{x\}}(\Omega^p_X) = 0$ for $p+q \leq n-1$ by (8.17), it follows that the differential $\delta : H^q_{\{x\}}(d\Omega^p_X) \to H^{q+1}_{\{x\}}(d\Omega^{p-1}_X)$ of the cohomology sequence of S^p is an injection when $p < m$, $p+q \leq n-1$. Thus we find a chain of injections

$$H^1_{\{x\}}(d\Omega^{m-1}_X) \hookrightarrow \ldots \hookrightarrow H^m_{\{x\}}(d\mathcal{O}_X) \hookrightarrow H^{m+1}_{\{x\}}(\mathbb{C}_X).$$

If $j : X-\{x\} \subset X$ denotes the inclusion, then the natural map $\mathrm{Ker}(d : \Omega^m_X \to \Omega^{m+1}_X) \to j_* j^{-1} d\Omega^{m-1}_X$ is an injection, since $H^0_{\{x\}}(\Omega^m_X) = 0$. This induces an injection $H^m(\Omega^{\cdot}_{X,x}) \hookrightarrow H^1_{\{x\}}(d\Omega^{m-1}_X)$, which composed with the injections above yields an injection $DR : H^m(\Omega^{\cdot}_{X,x}) \hookrightarrow H^{m+1}_{\{x\}}(\mathbb{C}_X)$.

This map interprets De Rham cohomology as ordinary cohomology. To be more precise, let B_ε be a closed ball about x in an ambient vector space of X such that the pair $(B_\varepsilon, B_\varepsilon \cap X)$ is isomorphic (in the sense of prop. (2.4)) to the cone on its boundary $(\partial B_\varepsilon, \partial B_\varepsilon \cap X)$. If $\omega \in H^m(\Omega^{\cdot}_{X,x})$, $m \geq 1$, then take ε so small that ω can be represented by a holomorphic m-form $\tilde{\omega}$ on B_ε whose restriction to $X-\{x\}$ is closed. Then $DR(\omega)$ is just the De Rham class of $\tilde{\omega}|B_\varepsilon \cap X-\{x\}$ in $H^m(B_\varepsilon \cap X-\{x\};\mathbb{C}) \cong H^{m+1}(B_\varepsilon \cap X, B_\varepsilon \cap X-\{x\}) \cong H^{m+1}_{\{x\}}(\mathbb{C}_X)$. We use this to show that DR is the zero-map: any m-cycle γ on $B_\varepsilon \cap X-\{x\}$ is homologous to an m-cycle γ' on $\partial B_\varepsilon \cap X$. The conical structure on $B_\varepsilon \cap X$ determines an $(m+1)$-cycle Γ on B_ε whose boundary is γ'. As $t.\Gamma$ $(0 < t \leq 1)$ is homologous to Γ, we have by Stokes,

$$\int_{\gamma'} \tilde{\omega} = \int_\Gamma d\tilde{\omega} = \int_{t.\Gamma} d\tilde{\omega}.$$

The last integral goes to zero as $t \to 0$, so $\int_{\gamma'} \tilde{\omega} = 0$. This proves that $DR(\omega) = 0$. Since DR is an injection it follows that $H^m(\Omega^{\cdot}_{X,x}) = 0$, in other words, S^m is exact. With induction on m, all the assertions of the proposition follow except the vanishing of $H^0_{\{x\}}(d\Omega^{n-1}_X)$. The exactness of S^{n-1} gives an inclusion $H^0_{\{x\}}(d\Omega^{n-1}_X) \hookrightarrow H^1_{\{x\}}(d\Omega^{n-2}_X)$. Composing it with the injection of the latter into $H^n_{\{x\}}(\mathbb{C}_X)$ gives a De Rham map

$DR : H^0_{\{x\}}(d\Omega^{n-1}_X) \to H^n_{\{x\}}(\mathbb{C}_X)$ which is interpreted as follows: any $\eta \in H^0_{\{x\}}(d\Omega^{n-1}_X)$ can be represented by an exact holomorphic n-form $d\tilde{\omega}$ on B_ε (ε sufficiently small) whose support avoids $(B_\varepsilon - \{x\}) \cap X$. Its value on an (n-1)-cycle γ on $\partial B_\varepsilon \cap X$ will be given by $\int_{\mathrm{cone}(\gamma)} d\tilde{\omega} = 0$. So $\eta = 0$ and (hence) $H^0_{\{x\}}(d\Omega^{n-1}_X) = 0$.

We use the preceding proposition to prove similar statements in the relative case. If $f : (X,x) \to (\mathbb{C}^k,0)$ is a deformation of an icis of dim n, then by (8.8) $H^n(\Omega^{\cdot}_{f,x})$ is a noetherian $\mathcal{O}_{\mathbb{C}^k,0}$-module. In view of the exactness of the sequences

$$H^n(\Omega^{\cdot}_{f,x}) \to \Omega^n_{f,x}/d\Omega^{n-1}_{f,x} \xrightarrow{d_f} \Omega^{n+1}_{f,x} \qquad \text{and}$$

$$\Omega^n_{f,x}/d\Omega^{n-1}_{f,x} \to \omega_{f,x}/d\Omega^{n-1}_{f,x} \to \omega_{f,x} \otimes \mathcal{O}_{C_f,x}$$

it follows that both $\Omega^n_{f,x}/d\Omega^{n-1}_{f,x}$ and $\omega_{f,x}/d\Omega^{n-1}_{f,x}$ are noetherian $\mathcal{O}_{\mathbb{C}^k,0}$-modules of the same rank as $H^n(\Omega^{\cdot}_{f,x})$. It also follows from (8.8) that this common rank equals the Milnor number of (X_0,x).

(8.20) *Proposition*. Let $f : (X,x) \to (\mathbb{C}^k,0)$ be a deformation of an icis of dim $n > 0$. Then $\Omega^n_{f,x}/d\Omega^{n-1}_{f,x}$ and $\omega_{f,x}/d\Omega^{n-1}_{f,x}$ are free $\mathcal{O}_{\mathbb{C}^k,0}$-modules of rank $\mu(X_0,x)$ and if f is a smoothing of (X_0,x) then the complex

$$0 \to \mathcal{O}_{\mathbb{C}^k,0} \xrightarrow{f^*} \mathcal{O}_{X,x} \xrightarrow{d_f} \Omega^1_{f,x} \to \dots \to \Omega^n_{f,x}$$

is exact.

Proof. We prove all the statements with induction on k under the assumption that f is a smoothing. This is sufficient, since $\Omega^n_{f,x}/d\Omega^{n-1}_{f,x}$ and $\omega_{f,x}/d\Omega^{n-1}_{f,x}$ are compatible with base change. After a coordinate change in $\mathbb{C}^k$ we may suppose that the coordinate hyperplane $\mathbb{C}^{k-1}\times\{0\}$ meets the discriminant of f in a germ of dim $\leq k-2$ (if $k \geq 2$). We put $X' = f_k^{-1}(0)$

and set $f'=(f_1,\dots,f_{k-1}) : (X',x) \to (\mathbb{C}^{k-1},0)$. For $p \le n$, $\Omega^p_{f,x}$ is torsion free and so

$$S^p : 0 \to \Omega^p_{f,x} \xrightarrow{t_k\cdot} \Omega^p_{f,x} \to \Omega^p_{f',x} \to 0$$

is exact. This gives rise to a cohomology sequence

$$0 \to H^0(\Omega^{\cdot}_{f,x}) \xrightarrow{t_k\cdot} H^0(\Omega^{\cdot}_{f,x}) \to H^0(\Omega^{\cdot}_{f',x}) \to H^1(\Omega^{\cdot}_{f,x}) \to \dots$$

$$\begin{array}{ccc} \cong & \cong & \cong \\ \mathcal{O}_{\mathbb{C}^k,0} & \mathcal{O}_{\mathbb{C}^k,0} & \mathcal{O}_{\mathbb{C}^{k-1},0} \end{array}$$

$$\dots \to H^{n-1}(\Omega^{\cdot}_{f',x}) \to \Omega^n_{f,x}/d\Omega^{n-1}_{f,x} \xrightarrow{t_k\cdot} \Omega^n_{f,x}/d\Omega^{n-1}_{f,x} \to \Omega^n_{f',x}/d\Omega^{n-1}_{f',x} \to 0$$

In case $k = 1$ it follows from (8.19) and in case $k \ge 2$ from our induction hypothesis that $H^p(\Omega^{\cdot}_{f',x}) = 0$ for $1 \le p \le n-1$. This implies that for $1 \le p \le n-1$, $t_k\cdot$ induces an isomorphism on $H^p(\Omega^{\cdot}_{f,x})$. Since $H^p(\Omega^{\cdot}_{f,x})$ is a noetherian $\mathcal{O}_{\mathbb{C}^k,0}$-module, Nakayama's lemma applied to the equality $t_kH^p(\Omega^{\cdot}_{f,x}) = H^p(\Omega^{\cdot}_{f,x})$ gives $H^p(\Omega^{\cdot}_{f,x}) = 0$. We have an exact sequence of noetherian $\mathcal{O}_{\mathbb{C}^k,0}$-modules

$$0 \to \Omega^n_{f,x}/d\Omega^{n-1}_{f,x} \xrightarrow{t_k\cdot} \Omega^n_{f,x}/d\Omega^{n-1}_{f,x} \to \Omega^n_{f',x}/d\Omega^{n-1}_{f',x} \to 0$$

By induction hypothesis, $\Omega^n_{f',x}/d\Omega^{n-1}_{f',x}$ is a free $\mathcal{O}_{\mathbb{C}^{k-1},0}$-module. This implies that $\Omega^n_{f,x}/d\Omega^{n-1}_{f,x}$ is a free $\mathcal{O}_{\mathbb{C}^k,0}$-module (let $\eta_1,\dots,\eta_\ell \in \Omega^n_{f,x}/d\Omega^{n-1}_{f,x}$ be a lift of an $\mathcal{O}_{\mathbb{C}^{k-1},0}$-basis of $\Omega^n_{f',x}/d\Omega^{n-1}_{f',x}$. Then $\eta_1,\dots,\eta_\ell$ generate $\Omega^n_{f,x}/d\Omega^{n-1}_{f,x}$ over $\mathcal{O}_{\mathbb{C}^k,0}$ by Nakayama's lemma. If $\phi_1\eta_1+\dots+\phi_\ell\eta_\ell = 0$, $\phi_\lambda \in \mathcal{O}_{\mathbb{C}^k,0}$, is a nontrivial relation, then divide it by the highest possible power of t_k to ensure that not all ϕ_λ belong to $t_k\mathcal{O}_{\mathbb{C}^k,0}$. But reducing the equation mod t_k implies $\phi_\lambda \in t_k\mathcal{O}_{\mathbb{C}^k,0}$, $\lambda = 1,\dots,\ell$. Contradiction.) The proof that $\omega_{f,x}/d\Omega^{n-1}_{f,x}$ is a free $\mathcal{O}_{\mathbb{C}^k,0}$-module is similar.

(8.21) *Corollary*. Let $f : X \to S$ be a good Stein representative of a smoothing $f : (X,x) \to (C^k,0)$ of an icis of dim $n > 0$. Then after possibly shrinking S, $H^p(f_*\Omega_f^\cdot) = 0$ for $0 < p < n$ and $H^n(f_*\Omega_f^\cdot)$ is a free $\mathcal{O}_S$-module of rank $\mu(X_0,x)$ fitting in the exact sequence

$$0 \to (R^n f_* \mathbb{C}_X) \otimes_{\mathbb{C}} \mathcal{O}_S \xrightarrow{\alpha^n \otimes 1} H^n(f_*\Omega_f^\cdot) \xrightarrow{\beta^n} f_* H^n(\Omega_f^\cdot) \to 0$$

where $\alpha^n \otimes 1$ is a section of the De Rham evaluation map.

Proof. By (8.6), the stalk of $H^p(f_*\Omega_f^\cdot)$ in 0 maps isomorphically to $H^p(\Omega_{f,x}^\cdot)$. So it follows from (8.20) and the fact that $H^p(f_*\Omega_f^\cdot)$ is coherent, that for a sufficiently small S, $H^p(f_*\Omega_f^\cdot)$ is trivial for $0 < p < n$ and is a free $\mathcal{O}_S$-module of rank $\mu(X_0,x)$ for $p = n$. Moreover, $f^{-1}\mathcal{O}_S \to H^0(\Omega_f^\cdot)$ will be an isomorphism (for S sufficiently small) and so by lemma (8.1), $R^p f_* H^0(\Omega_f^\cdot)$ may be identified with $(R^p f_* \mathbb{C}_X) \otimes_{\mathbb{C}} \mathcal{O}_S$. The exact sequence now follows from (8.5).

(8.22) *Corollary*. For an icis (X_0,x) of dim $n > 0$ we have a natural exact sequence of finite dimensional $\mathbb{C}$-vector spaces

$$0 \to H^0_{\{x\}}(\Omega^n_{X_0,x}) \to \Omega^n_{X_0,x}/d\Omega^{n-1}_{X_0,x} \to \omega_{X_0,x}/d\Omega^{n-1}_{X_0,x} \to \omega_{X_0,x}/\Omega^n_{X_0,x} \to 0$$

of which the middle two terms have $\mathbb{C}$-dimension $\mu(X_0,x)$.

Proof. The sequence is a quotient of the stalk of the sequence (8.18a) in x. Since $d\Omega^{n-1}_{X_0,x} \cap H^0_{\{x\}}(\Omega^n_{X_0,x}) = H^0_{\{x\}}(d\Omega^{n-1}_{X_0,x}) = 0$ by (8.19), the exactness is preserved. The assertion concerning the Milnor number follows from (8.8) and (8.20).

(8.23) Notice that the corollary implies that $H^0_{\{x\}}(\Omega^n_{X_0,x})$ and $\omega_{X_0,x}/\Omega^n_{X_0,x}$ have the same $\mathbb{C}$-dimension. If (X_0,x) is given by a germ $f : (\mathbb{C}^{n+k},x) \to (\mathbb{C}^k,0)$, then exterior multiplication with $df_1 \wedge \ldots \wedge df_k$

identifies $\omega_{X_0,x}/\Omega^n_{X_0,x}$ with the Artin ring

$$\mathcal{O}_{\mathbb{C}^{n+k},x}/(\{\frac{\partial(f_1,\ldots,f_k)}{\partial(z_{i_1},\ldots,z_{i_k})} : 1\le i_1<\ldots<i_k\le n+k\},f_1,\ldots,f_k)\mathcal{O}_{\mathbb{C}^{n+k},x}$$

and so its dimension is relatively easy to compute. We denote this number by $\tau'(X_0,x)$. From the preceding corollary it is clear that $\tau'(X_0,x) \le \mu(X_0,x)$. For hypersurface singularities it is almost immediate that $\tau'(X_0,x) = \tau(X_0,x)$. Greuel (1980) asks whether it is always true that $\tau'(X_0,x) \le \mu(X_0,x)$.

Let $f : (X,x) \to (\mathbb{C}^k,0)$ be a smoothing of an icis of dim n. We defined a local Gauss-Manin connection ∇ as an integrable connection on $H^n(\Omega^{\cdot}_{f,x})[\delta^{-1}]$, where δ is an equation for the discriminant of f. Both for theoretical and practical purposes it is often convenient to describe ∇ in terms of certain $\mathcal{O}_{\mathbb{C}^k,0}$-submodules of $H^n(\Omega^{\cdot}_{f,x})[\delta^{-1}]$ which generate the latter over $\mathcal{O}_{\mathbb{C}^k,0}[\delta^{-1}]$. We have for instance:

(8.24) *Proposition.* There is a natural chain of inclusions of $\mathcal{O}_{\mathbb{C}^k,0}$-modules

$$H^n(\Omega^{\cdot}_{f,x}) \subset \Omega^n_{f,x}/d\Omega^{n-1}_{f,x} \subset \omega_{f,x}/d\Omega^{n-1}_{f,x} \subset H^n(\Omega^{\cdot}_{f,x})[\delta^{-1}]$$

and if $\eta \in \theta_{\mathbb{C}^k,0}$, resp. is a liftable element of $\theta_{\mathbb{C}^k,0}$ then ∇_η maps each of these modules in the next resp. to itself.

Proof. The cokernel of the first resp. second inclusion is contained in $\Omega^{n+1}_{f,x}$ resp. equals $\omega_{f,x}/\Omega^n_{f,x}$. It is clear that both are supported by the critical locus of f and are therefore killed by some power of δ. On the other hand, $\omega_{f,x}/d\Omega^{n-1}_{f,x}$ is a free $\mathcal{O}_{\mathbb{C}^k,0}$-module by (8.20), so that it indeed injects into $H^n(\Omega^{\cdot}_{f,x})[\delta^{-1}]$.

For the second part of the proposition we use formula (8.9a):

if $\omega \in \Omega^n_{X,x}$ represents $[\omega] \in H^n(\Omega^{\cdot}_{f,x})$, then $d\omega = \Sigma_\kappa df_\kappa \wedge \omega_\kappa$ for certain $\omega_\kappa \in \Omega^n_{X,x}$ and so $\nabla([\omega]) = \Sigma_\kappa\, dt_\kappa \otimes [\omega_\kappa]$ with each $[\omega_\kappa] \in \Omega^n_{f,x}/d\Omega^{n-1}_{f,x}$.

If $\omega \in \Omega^n_{X,x}$ represents $[\omega] \in \Omega^n_{f,x}/d\Omega^{n-1}_{f,x}$, then

$$df_1\wedge\ldots\wedge df_k \wedge \nabla([\omega]) = \Sigma_\kappa\, dt_\kappa \otimes (-1)^{k-\kappa} df_1\wedge\ldots\widehat{df_\kappa}\ldots\wedge df_k\wedge d\omega$$

(for this is true if $[\omega] \in H^n(\Omega^{\cdot}_{f,x})$), which shows that $\nabla_{\partial/\partial t_\kappa}([\omega])$ is in $\omega_{f,x}/d\Omega^{n-1}_{f,x}$.

Finally, if $\xi \in \theta_{X,x}$ lifts $\eta \in \theta_{\mathbb{C}^k,0}$, then L_ξ induces endomorphisms of $\Omega^n_{f,x}$ and $\omega_{f,x}$ (the proof can safely be left to the reader), so that ∇_η preserves the modules in question.

*(8.25) If $n \geq 2$, then $\omega_{f,x} = j_*j^{-1}\Omega^n_{f,x}$, where $j : X-C \subset X$ $(C = C_f)$, so that $\omega_{f,x}/d\Omega^{n-1}_{f,x}$ may be identified with $H^1_C(d\Omega^{n-1}_{f,x})$. This module is at the beginning of another chain of inclusions

$$\begin{array}{ccccccc} H^1_C(d\Omega^{n-1}_{f,x}) & \hookrightarrow H^2_C(d\Omega^{n-2}_{f,x}) & \hookrightarrow \ldots \hookrightarrow & H^n_C(d\mathcal{O}_{X,x}) & \hookrightarrow H^{n+1}_C(f^{-1}\mathcal{O}_{\mathbb{C}^k,0}) \\ \shortparallel & \shortparallel & & \shortparallel & \\ \mathcal{F}^n & \mathcal{F}^{n-1} & & \mathcal{F}^1 & \end{array}$$

and $\mathcal{F}^p/\mathcal{F}^{p+1}$ can be identified with a submodule of $H^{n+1-p}_C(\Omega^p_{f,x})$. This follows from the exactness of

$$0 \to d\Omega^{p-1}_{f,x} \to \Omega^p_{f,x} \to d\Omega^p_{f,x} \to 0 \qquad (0 \leq p \leq n)$$

and the vanishing assertions (8.17). For $p > 0$, $H^{n+1-p}_C(\Omega^p_{f,x})$ *is a* noetherian $\mathcal{O}_{X,x}$-module supported by (C,x), so that $\mathcal{F}^1,\ldots,\mathcal{F}^n$ are noetherian $\mathcal{O}_{\mathbb{C}^k,0}$-modules. We can identify $\mathcal{F}^p$ with the set of relative cohomology classes in $H^{n+1}_C(f^{-1}\mathcal{O}_{\mathbb{C}^k,0})$ which are representable by a closed relative C^∞ n-form ω on the intersection of a neighbourhood U of x with X-C such that ω is a sum of forms of type $(p',n-p')$ with $p' \geq p$. Furthermore, for any

$\eta \in \theta_{\mathbb{C}^k,0}$, ∇_η maps each of these modules to the next and the induced maps

$$\begin{array}{ccc} \mathrm{Def}(\nabla_\eta) : F^p/F^{p+1} & \to & F^{p-1}/F^p \\ \cap & & \cap \\ H_C^{n+1-p}(\Omega^p_{f,x}) & & H_C^{n+2-p}(\Omega^{p-1}_{f,x}) \end{array}$$

are simply obtained by taking the cup product with a so-called Kodaira-Spencer class $\overline{\rho}_f(\eta) \in H^2_C(\theta_{f,x})$, which we do not define (suffice to say here that it only depends on $\rho_f(\eta) \in T_{f,x}$ — whence the name). This is an analogue of Griffith's transversality.

(8.26) Let us now take a closer look at the hypersurface case: $f : (\mathbb{C}^{n+1},0) \to (\mathbb{C},0)$. We identify $\omega_{f,0}$ with $\mathcal{O}_{\mathbb{C}^{n+1},0}$ by means of: $\alpha \mapsto$ coefficient of $df \wedge \alpha$ with respect to $dz_0 \wedge \ldots \wedge dz_n$. This makes $d\Omega^{n-1}_{f,0}$ correspond to the $\phi \in \mathcal{O}_{\mathbb{C}^{n+1},0}$ which are of the form $\Sigma^n_{\nu=0}\, \eta_\nu \frac{\partial f}{\partial z_\nu}$ with $\Sigma^n_{\nu=0} \frac{\partial \eta_\nu}{\partial z_\nu} \in (\frac{\partial f}{\partial z_0}, \ldots, \frac{\partial f}{\partial z_n})\mathcal{O}_{\mathbb{C}^{n+1},0}$. This is a $\mathbb{C}\{f\}$-submodule of $\mathcal{O}_{\mathbb{C}^{n+1},0}$ - as it ought to be - which we denote by $M(f)$. According to (8.20), $\mathcal{O}_{\mathbb{C}^{n+1},0}/M(f)$ is a free $\mathbb{C}\{f\}$-module of rank μ, so that $\mathcal{O}_{\mathbb{C}^{n+1},0}/M(f) + f\mathcal{O}_{\mathbb{C}^{n+1},0})$ is a $\mathbb{C}$-vector space of dim μ. In practice it is not so easy to find μ elements in $\mathcal{O}_{\mathbb{C}^{n+1},0}$ representing a basis of this vector space, although there is an algorithm, see Brieskorn (1970a), Scherk (1980).

Things are considerably simpler if f is weighted homogeneous. Recall from (1.4) that this means there are positive integers $d_0,\ldots,d_n,N$ such that f is a $\mathbb{C}$-linear combination of monomials $z_0^{i_0}\ldots z_n^{i_n}$ with $d_0 i_0 + \ldots + d_n i_n = N$, or what amounts to the same, that the derivation $\xi := \Sigma_\nu\, c_\nu z_\nu\, \partial/\partial z_\nu$, with $c_\nu := N^{-1} d_\nu$, leaves f invariant. So ξ is then a lift of $t\frac{d}{dt}$. Since $f \in (\frac{\partial f}{\partial z_0}, \ldots, \frac{\partial f}{\partial z_n})\mathcal{O}_{\mathbb{C}^{n+1},0}$, it follows that

$M(f) + f\mathcal{O}_{\mathbb{C}^{n+1},0} \subset (\frac{\partial f}{\partial z_0}, \ldots, \frac{\partial f}{\partial z_n})\mathcal{O}_{\mathbb{C}^{n+1},0}$. As both have the same

$\mathbb{C}$-codimension in $\mathcal{O}_{\mathbb{C}^{n+1},0}$ (namely μ), the two subspaces must be equal. Let $\phi_1,\ldots,\phi_\mu \in \mathbb{C}[z_0,\ldots,z_n]$ be monomials which project onto a $\mathbb{C}$-basis of $\mathcal{O}_{\mathbb{C}^{n+1},0}/(\frac{\partial f}{\partial z_0},\ldots,\frac{\partial f}{\partial z_n})\mathcal{O}_{\mathbb{C}^{n+1},0}$. Then they project also on a $\mathbb{C}\{t\}$-basis $[\phi_1],\ldots,[\phi_\mu]$ of $\mathcal{O}_{\mathbb{C}^{n+1},0}/M(f)$. If we put $\deg(z_\nu) = c_\nu$, then it is easily verified that L_ξ multiplies $\phi_\alpha\, dz_0 \wedge\ldots\wedge dz_n/df$ with $\deg(\phi_\alpha)+c_0+\ldots+c_n-1$. We put $r := c_0+\ldots+c_n-1$. A horizontal (multivalued) section of $\mathcal{O}_{\mathbb{C}^{n+1},0}/M(f)$ $(\cong \omega_{f,0}/d\Omega^{n-1}_{f,0})$ is then given by

$$t \to t^{-\deg(\phi_\alpha)-r}[\phi_\alpha]$$

If we substitute $t = \rho e^{2\pi i\theta}$ and let θ go from 0 to 1, we find that the monodromy operator multiplies $[\phi_\alpha]$ with $e^{-2\pi i(\deg\phi_\alpha+r)}$. The monodromy on cohomology is the inverse dual of the monodromy on homology. So we find that the latter is diagonalizable over $\mathbb{C}$ and has characteristic polynomial

$$\Pi^{\mu}_{\alpha=1}\ (T - e^{2\pi i(\deg(\phi_\alpha)+r)}).$$

9 APPLICATIONS OF THE LOCAL GAUSS-MANIN CONNECTION

We discuss three applications of the results of the preceding chapter which in some way are all related to each other. We begin with defining a so-called period mapping for the miniversal deformation of an icis (X_0,x). This mapping has the complement of the discriminant (or rather a covering thereof) as its source and the complex cohomology group of a Milnor fibre as its target. We prove that under certain conditions, this map is a local immersion. This implies that then $\tau(X_0,x) \leq \mu(X_0,x)$ and we thus recover a result of Greuel. The second section begins with discussing isolated singularities with good $\mathbb{C}^*$-action in general. In the complete intersection case, we define a certain pairing which we prove to be perfect. This enables us to conclude that $\tau(X_0,x) = \mu(X_0,x)$ in that case ($\dim(X_0,x) \geq 2$), a result which is also due to Greuel. In the final section we use the period mapping to investigate the miniversal deformation of a Kleinian singularity. In particular, we obtain an identification of the discriminant of its miniversal deformation with the discriminant of its associated Coxeter group.

9.A *Milnor number and Tjurina number*

(9.1) We begin with recalling some elementary linear algebra. Let

V be a finite dimensional vector space over a field k. Any $v \in V$ defines a *contraction* $\iota_v : V^* \to k$, $\phi \mapsto \phi(v)$. This contraction extends uniquely to an anti-derivation of degree -1 of the exterior algebra $\Lambda^\cdot V^*$ of V (i.e. ι_v sends $\Lambda^p V^*$ to $\Lambda^{p-1}V^*$ and $\iota_v(\alpha\wedge\beta) = \iota_v(\alpha)\wedge\beta + (-1)^{\deg\alpha}\alpha\wedge\iota_v(\beta)$). For any $\phi \in V^*$ we have

$$\iota_v(\phi\wedge\alpha) + \phi\wedge\iota_v(\alpha) = \phi(v)\alpha$$

(this is clearly true if $\alpha \in V^*$; since $\alpha \to \iota_v(\phi\wedge\alpha) + \phi\wedge\iota_v(\alpha)$ is a derivation of degree 0, the validity for all $\alpha \in \Lambda^\cdot V^*$ is now a formal consequence). So if $\phi(v) = 1$, then $\phi\wedge$ resp. ι_v provide chain homotopies in the complexes

$$\ldots \to \Lambda^{p+1}V^* \xrightarrow{\iota_v} \Lambda^p V^* \to \ldots \to V^* \xrightarrow{\iota_v} k \to 0$$

$$0 \to k \xrightarrow{\phi\wedge} V^* \xrightarrow{\phi\wedge} \Lambda^2 V^* \to \ldots \to \Lambda^p V^* \xrightarrow{\phi\wedge} \Lambda^{p+1}V^* \to \ldots,$$

proving their exactness. If we put parameters in this argument we get corresponding assertions for (germs of) vector bundles. We shall use these observations without explicit reference.

The following lemma, due to Schlessinger (1971), is useful when dealing with modules of derivations.

(9.2) *Lemma*. Let (X,x) be an analytic germ of depth ≥ 2. Then for any noetherian $\mathcal{O}_{X,x}$-module M, the dual $\mathrm{Hom}_{\mathcal{O}_{X,x}}(M,\mathcal{O}_{X,x})$ has also depth ≥ 2.
Proof. Choose a presentation

$$\mathcal{O}^q_{X,x} \xrightarrow{\beta} \mathcal{O}^p_{X,x} \xrightarrow{\alpha} M \to 0$$

of M. Applying $\mathrm{Hom}_{\mathcal{O}_{X,x}}(-,\mathcal{O}_{X,x})$ to it gives an exact sequence

$$0 \to \mathrm{Hom}_{\mathcal{O}_{X,x}}(M,\mathcal{O}_{X,x}) \xrightarrow{\alpha^*} \mathcal{O}^p_{X,x} \xrightarrow{\beta^*} \beta^*(\mathcal{O}^p_{X,x}) \to 0$$
$$\beta^*(\mathcal{O}^p_{X,x}) \subset \mathcal{O}^q_{X,x}$$

By assumption $H^i_{\{x\}}(\mathcal{O}_{X,x}) = 0$ for $i = 0,1$ and so the associated local cohomology sequence gives $H^i_{\{x\}}(\mathrm{Hom}_{\mathcal{O}_{X,x}}(M,\mathcal{O}_{X,x})) = 0$ for $i = 0,1$.

We first aim to prove

(9.3) *Proposition.* Let $f : (X,x) \to (S,0)$ be a miniversal deformation of an icis (X_0,x) of dim $n \geq 1$ and let $\omega_0 \in \omega_{X_0,x}$ be a generator of the dualizing module. Denote (as before) by $\theta_{S,0}\langle D\rangle$ the submodule of derivations in $\theta_{S,0}$ preserving the discriminant. Then there is a well-defined map

$$\tilde{\iota} : \theta_{S,0}\langle D\rangle/m_{S,0}\theta_{S,0}\langle D\rangle \to H^1_{\{x\}}(\Omega^{n-1}_{X_0,x})$$

given by $\eta \mapsto (\iota_{\tilde{\eta}}|X_0)(\omega_0)$, where $\tilde{\eta} \in \theta_{X,x}$ is a lift of η of f. This map is injective; if $n \geq 2$ it is also surjective, in particular, $H^1_{\{x\}}(\Omega^{n-1}_{X_0,x})$ has then $\mathbb{C}$-dimension $\tau(X_0,x)$.

We do this via three lemmas

(9.4) *Lemma.* Let $\tilde{\theta}_{f,x} \subset \theta_{X,x}$ denote the $\mathcal{O}_{S,0}$-module of derivations in $\theta_{X,x}$ lifting a derivation in $\theta_{S,0}$. Then the natural map

$$\tilde{\theta}_{f,x} \otimes_{\mathcal{O}_{S,0}} \mathbb{C} \to \theta_{X,x} \otimes_{\mathcal{O}_{S,0}} \mathbb{C}$$

is injective (this is equivalent to $\tilde{\theta}_{f,x}$ being flat over $\mathcal{O}_{S,0}$).

Proof. In view of the exact sequence

$$0 \to \tilde{\theta}_{f,x} \to \theta_{X,x} \to \theta(f)_x/f^*(\theta_{S,0}) \to 0$$

it suffices to show that $\theta(f)_x/f^*(\theta_{S,0})$ is flat over $\mathcal{O}_{S,0}$. This module is isomorphic to a direct sum of copies of $\mathcal{O}_{X,x}/f^{-1}(\mathcal{O}_{S,0})$, so we must prove that $\mathcal{O}_{X,x}/f^{-1}(\mathcal{O}_{S,0})$ is flat over $\mathcal{O}_{S,0}$. Consider the exact sequence of $\mathcal{O}_{S,0}$-modules

$$0 \to \mathcal{O}_{S,0} \overset{\phi}{\to} \mathcal{O}_{X,x} \overset{\psi}{\to} \mathcal{O}_{X,x}/f^{-1}(\mathcal{O}_{S,0}) \to 0$$

The local ring $\mathcal{O}_{X,x}$ is flat over $\mathcal{O}_{S,0}$, hence faithfully flat (Matsumura (1980), Ch. 2, Thm 3). This implies that for any noetherian $\mathcal{O}_{S,0}$-module N, $\phi \otimes \mathbf{1}_N$ is injective (*loc.cit.*, Thm. 2(i)). Since $\mathrm{Tor}_1^{\mathcal{O}_{S,0}}(\mathcal{O}_{X,x},N) = 0$, it follows that $\mathrm{Tor}_1^{\mathcal{O}_{S,0}}(\mathcal{O}_{X,x}/f^{-1}(\mathcal{O}_{S,0}),N) = 0$. This just means that $\mathcal{O}_{X,x}/f^{-1}(\mathcal{O}_{S,0})$ is flat over $\mathcal{O}_{S,0}$.

(9.5) *Lemma*. The inclusion $r : \tilde{\theta}_{f,x} \subset \theta_{X,x}$ induces an isomorphism $\tilde{\theta}_{f,x} \otimes_{\mathcal{O}_{S,0}} \mathbb{C} \to \theta_{X_0,x}$.

Proof. By the preceding lemma, $r \otimes \mathbf{1}_{\mathbb{C}}$ is injective. So we have to show that its image equals $\theta_{X_0,x} \subset \theta_{X,x} \otimes_{\mathcal{O}_{S,s}} \mathbb{C}$.

If $\xi_0 \in \theta_{X_0,x}$, then let $\xi \in \theta_{X,x}$ be such that $\xi_0 = \xi \otimes \mathbf{1}$. Clearly, $\partial f(\xi) \in m_{S,0}\theta(f)_x$. As f is versal, we have

$\theta(f)_x = \partial f(\theta_{X,x}) + f^*(\theta_{S,0})$ and so there exist $\xi' \in m_{S,0}\theta_{X,0}$ and $\eta' \in m_{S,0}\theta_{S,0}$ with $\partial f(\xi) = \partial f(\xi') + f^*(\eta')$. Then $\xi-\xi' \in \tilde{\theta}_{f,x}$ and $(\xi-\xi') \otimes \mathbf{1} = \xi \otimes \mathbf{1} = \xi_0$. Hence $\theta_{X_0,x} \subset \mathrm{Im}(r \otimes \mathbf{1}_{\mathbb{C}})$.

To prove the opposite inclusion, we use the fact that the kernel of the Kodaira-Spencer map $\rho_f : \theta_{S,0} \to T_{f,x}$ is contained in $m_{S,0}\theta_{S,0}$ (for f is miniversal, see (6.5)). So if $\xi \in \tilde{\theta}_{f,x}$ lifts $\eta \in \theta_{S,0}$, then $\eta \in m_{S,0}\theta_{S,0}$. This implies that ξ preserves the ideal $f^*(m_{S,0})\mathcal{O}_{X,x}$ defining (X_0,x) and so $\xi \otimes \mathbf{1} \in \theta_{X_0,x}$.

Let $\xi_0 \in \theta_{X_0,x}$. In general $\iota_{\xi_0}(\omega_0)$ will be an $(n-1)$-form on a punctured neighbourhood of x in X_0.

(9.6) *Lemma*. We have an exact sequence

$$\theta_{f,x} \to \theta_{X_0,x} \xrightarrow{\iota} H^1_{\{x\}}(\Omega^{n-1}_{X_0,x})$$

where ι is defined by $\xi_0 \to \iota_{\xi_0}(\omega_0)$ and the first map is given by restriction. Moreover, ι is surjective if $n \geq 2$.

Proof. Lift ω_0 to a generator ω of $\omega_{f,x}$. If $\xi_0 \in \theta_{X_0,x}$ lifts to $\xi \in \theta_{f,x}$, then $\iota_{\xi_0}(\omega_0)$ is the restriction of $\iota_\xi(\omega)$. Now, $\iota_\xi(\omega)$ lives in $j_*j^{-1}\Omega^{n-1}_{f,x}$, where $j : X-C_f \subset X$, and hence extends to an element of $\Omega^{n-1}_{f,x}$ (because $H^1_{C_f}(\Omega^{n-1}_{f,x}) = 0$ by (8.17)). It follows that $\iota_{\xi_0}(\omega_0)$ comes from $\Omega^{n-1}_{f,x}$.

Conversely, if $\iota_{\xi_0}(\omega_0)$ lifts to $\zeta_0 \in \Omega^{n-1}_{X_0,x}$, then lift ζ_0 to $\zeta \in \Omega^{n-1}_{f,x}$. Since ω is a generator of Ω^n_f on a complement neighbourhood $U-(C_f \cap U)$ of x in X, there must be a unique $\xi \in j_*j^{-1}\theta_{f,x}$ such that $\iota_\xi\omega = \zeta$ on such a set. As $\operatorname{codim}_x C_f \geq 2$, we have $j_*j^{-1}\theta_{f,x} = \theta_{f,x}$ and so $\xi \in \theta_{f,x}$. If we denote the image of ξ in $\theta_{X_0,x}$ by ξ_0', then it is clear that $\xi_0-\xi_0' \in H^0_{\{x\}}(\theta_{X_0,x})$. The last group is trivial for $\theta_{X_0,x}$ is the $\mathcal{O}_{X_0,x}$-dual of an $\mathcal{O}_{X_0,x}$-module (namely $\Omega_{X_0,x}$) and hence torsion free. So $\xi_0 = \xi_0'$.

Finally, if ζ_0 is an $(n-1)$-form on a punctured neighbourhood of x in X_0, then there is a derivation ξ_0 on a punctured neighbourhood of x in X_0 with $\iota_{\xi_0}\omega_0 = \zeta_0$. If $n \geq 2$, then $H^1_{\{x\}}(\theta_{X_0,x}) = 0$ by (9.2) and so $\xi_0 \in \theta_{X_0,x}$. This proves that ι is surjective in this case.

Proof of (9.3). Recall from (6.13) and (6.14) that $\theta_{S,0}\langle D\rangle$ is a free $\mathcal{O}_{S,0}$-module (of rank $\tau(X_0,x)$) and consists of the $\eta \in \theta_{S,0}$ which lift over f. So we have an exact sequence of $\mathcal{O}_{S,0}$-modules

$$0 \to \theta_{f,x} \to \tilde{\theta}_{f,x} \xrightarrow{\partial f} \theta_{S,0}\langle D\rangle \to 0.$$

The proposition is now immediate from the two preceding lemmas.

One of the interesting aspects of $\tilde{\iota}$ is that for $n \geq 2$ it occurs as a factor in a 'punctual form' of the Gauss-Manin connection. To be precise, recall that any $\eta \in \theta_{S,0}\langle D\rangle$ acts on $\omega_{f,x}/d\Omega^{n-1}_{f,x}$ (as the covariant derivative) by lifting η to $\tilde{\eta} \in \tilde{\theta}_{f,x}$ and putting

$$\nabla_\eta([\omega]) := [L_{\tilde{\eta}}\omega] = [d\iota_{\tilde{\eta}}\omega], \qquad \omega \in \omega_{f,x}.$$

So the fibre of $\eta \in \theta_{S,0}\langle D\rangle \to \nabla_\eta([\omega])$ over $0 \in S$ is just the composite

$$\theta_{S,0}\langle D\rangle/m_{S,0}\theta_{S,0}\langle D\rangle \overset{\tilde{\iota}}{\hookrightarrow} H^1_{\{x\}}(\Omega^{n-1}_{X_0,x}) \overset{d}{\to} H^1_{\{x\}}(d\Omega^{n-1}_{X_0,x})$$

$$\cup$$

$$\omega_{X_0,x}/d\Omega^{n-1}_{X_0,x}$$

For $n \geq 2$ the two inclusions are isomorphisms, for $n = 1$ we only know that $d\circ\tilde{\iota}$ must map into $\omega_{X_0,x}/d\mathcal{O}_{X_0,x}$. The map d fits in an exact sequence associated to $0 \to d\Omega^{n-2}_{X_0} \to \Omega^{n-1}_{X_0} \to d\Omega^{n-1}_{X_0} \to 0$:

$$0 \to H^1_{\{x\}}(d\Omega^{n-2}_{X_0,x}) \to H^1_{\{x\}}(\Omega^{n-1}_{X_0,x}) \overset{d}{\to} H^1_{\{x\}}(d\Omega^{n-1}_{X_0,x})$$

(we have used here that $H^0_{\{x\}}(d\Omega^{n-1}_{X_0,x}) = 0$, by (8.19). For $n = 1$, $d\Omega^{n-2}_{X_0}$ must be interpreted as $\mathbb{C}_{X_0}$). For future reference we sum up:

(9.7) *Corollary*. Assume $n \geq 1$. Then

$$\theta_{S,0}\langle D\rangle \to \omega_{f,x}/d\Omega^{n-1}_{f,x}, \qquad \eta \mapsto \nabla_\eta([\omega])$$

is a homomorphism of free $\mathcal{O}_{S,0}$-module of rank $\tau(X_0,x)$ and $\mu(X_0,x)$ respectively. The kernel of the map on the fibres over $0 \in S$,

$$\theta_{S,0}\langle D\rangle/m_{S,0}\theta_{S,0}\langle D\rangle \to \omega_{X_0,x}/d\Omega^{n-1}_{X_0,x},$$

can be identified with $H^1_{\{x\}}(d\Omega^{n-2}_{X_0,x})$ for $n \geq 2$ and with a subspace of $H^1_{\{x\}}(\mathbb{C}_{X_0,x})$ if $n = 1$. In particular, if $H^1_{\{x\}}(d\Omega^{n-2}_{X_0,x}) = 0$ (a condition satisfied if X_0 is a rational homology sphere at x), then the above maps are injective and hence $\tau(X_0,x) \leq \mu(X_0,x)$.

The significance of this corollary will become more apparent when we discuss the period mapping in section C.

For $n = 1$, local duality gives a perfect pairing $H^0_{\{x\}}(\Omega^1_{X_0,x}) \times \mathrm{Ext}^1_{\mathcal{O}_{X_0,x}}(\Omega^1_{X_0,x},\mathcal{O}_{X_0,x}) \to \mathbb{C}$ (see e.g. Greuel (1980)), and hence $\tau(X_0,x) = \tau'(X_0,x)$ (notation of (8.23)), so that then $\tau(X_0,x) \leq \mu(X_0,x)$ always. The inequality $\tau(X_0,x) \leq \mu(X_0,x)$ holds in fact for all $n \geq 1$ (Steenbrink & Looijenga (1983)). Our next aim is to prove that we have equality: $\tau(X_0,x) = \mu(X_0,x)$ if (X_0,x) has 'good $\mathbb{C}^*$-action'.

9.B *Singularities with good* $\mathbb{C}^*$*-action*

As usual $\mathbb{C}^*$ denotes the multiplication group of $\mathbb{C}$. If V is a representation, then $v \in V$ is called an *eigenvector* if $\mathbb{C}^*$ leaves the line $\mathbb{C}v$ invariant. The $\mathbb{C}^*$-action on this line is then simply given by $\lambda.v = \lambda^d v$ for some (unique) integer $d \in \mathbb{Z}$ called the *weight* of v. A basic property of $\mathbb{C}^*$ is that it is semi-simple: any finite dimensional representation is a direct sum of eigenvectors. The set of weights of these eigenvectors (counted with multiplicities) is an invariant of the representation (in fact characterizes it up to isomorphism).

It is clear what we will mean by a $\mathbb{C}^*$-action on a complex-analytic germ (X,x): this is just a group homomorphism $\mathbb{C}^* \to \mathrm{Aut}(X,x)$ such that the evaluation map-germ $\mathbb{C}^*\times(X,x) \to (X,x)$ is analytic. This determines a representation of $\mathbb{C}^*$ on the $\mathbb{C}$-algebra $\mathcal{O}_{X,x}$. We say that the $\mathbb{C}^*$-action is *good* if $\mathbb{C}^*$ has only positive weights in $m_{X,x}$.

(9.8) *Lemma*. Let (X,x) be a reduced analytic germ endowed with a $\mathbb{C}^*$-action. Then the action is good if and only if the weights of $m_{X,x}/m_{X,x}^2$ are all > 0. In the case of a good $\mathbb{C}^*$-action, the subspace $R \subset \mathcal{O}_{X,x}$ spanned by the eigenvectors is a noetherian $\mathbb{C}$-subalgebra, graded by its weights: $R = \bigoplus_{d=0}^{\infty} R_d$ with the property that the natural map from $\hat{R} := \prod_{d=0}^{\infty} R_d$ to the $m_{X,x}$-adic completion of $\mathcal{O}_{X,x}$ is an isomorphism.

Proof. We first show that eigenvectors in $m_{X,x}^k/m_{X,x}^{k+1}$ can be lifted to eigenvectors in $m_{X,x}^k$. If $\phi \in m_{X,x}^k$ represents an eigenvector of $m_{X,x}^k/m_{X,x}^{k+1}$ of weight w, then choose a neighbourhood U of x in X on which ϕ converges and which is invariant under the compact group $S^1 = \{\lambda \in \mathbb{C}^*: |\lambda| = 1\}$. Consider

$$\phi'(z) := \int_0^1 e^{-2\pi i w\theta}(e^{2\pi i\theta}.\phi)(z)\,d\theta$$

It is easily checked that $\phi'(\lambda.z) = \lambda^w\phi'(z)$ if $|\lambda| = 1$. If $\phi'(z) \neq 0$, then $\lambda \to \phi'(\lambda.z)/(\lambda^w\phi'(z))$ is constant one on the unit circle and hence constant one on a neighbourhood of $1 \in \mathbb{C}^*$. It follows that ϕ' is an eigenvector of weight w. In the equality

$$\phi(z)-\phi'(z) = \int_0^1 (\phi-e^{-2\pi i w\theta}(e^{2\pi i\theta}.\phi))(z)]d\theta$$

the integrand is for any fixed $\theta \in [0,1]$ in $m_{X,x}^{k+1}$ and so $\phi-\phi' \in m_{X,x}^{k+1}$.

In particular, we can find eigenvectors $\phi_1,\dots,\phi_N \in m_{X,x}$ mapping to a basis of $m_{X,x}/m_{X,x}^2$. If we denote their weights by $w_1,\dots,w_N$, then the weights of $m_{X,x}^k/m_{X,x}^{k+1}$ will be positive linear combinations of $w_1,\dots,w_N$. So the action is good if and only if all w_ν are positive.

Assume the action is good. We first show that $\phi_1,\dots,\phi_N$

generate R as a $\mathbb{C}$-algebra. It is clear that the monomials of degree k in the $\phi_1,\ldots,\phi_N$ map onto a generating set of $m^k_{X,x}/m^{k+1}_{X,x}$. As these are all eigenvectors, the weights of $m^k_{X,x}/m^{k+1}_{X,x}$ are $\geq kw_{min}$ where $w_{min} = \min\{w_1,\ldots,w_N\}$. If $\phi \in O_{X,x}$ is an eigenvector of weight w, then we can choose an eigenvector ψ of weight w which is a polynomial expression in the $\phi_1,\ldots,\phi_N$ such that $\phi-\psi \in m^k_{X,x}$ for some $k > w/w_{min}$. As $\phi-\psi$ has weight w, we must have $\phi-\psi \in m^\ell_{X,x}$ for all ℓ and so $\phi = \psi \in \mathbb{C}[\phi_1,\ldots,\phi_N]$. This argument also proves that for $k > w/w_{min}$ the natural maps

$$R \to O_{X,x}/m^k_{X,x} \cong \hat{O}_{X,x}/\hat{m}^k_{X,x} \leftarrow \hat{O}_{X,x}$$

are isomorphisms when restricted to the eigenspaces of weight w.

The geometric meaning of the last part of this lemma is that (X,x) may be viewed as the germ of an affine variety with good $\mathbb{C}^*$-action (namely Spec(R)) at its unique fixed point (defined by the maximal ideal $R_+ = \bigoplus_{d=1}^{\infty} R_d$). A hypersurface germ admits a good $\mathbb{C}^*$-action if and only if there exists a coordinate system in which it has a weighted homogeneous defining equation. So '(X,x) admits a good $\mathbb{C}^*$-action' is a coordinate invariant generalization of the expression '(X,x) is weighted homogeneous'. The normal surface singularities (X,x) with good $\mathbb{C}^*$-action are precisely the quasi-cones (1.D). These have been studied extensively by Orlik-Wagreich (1971), Dolgachev (1975), Pinkham (1977a), Wahl (1982, 1983), Neumann (1983) and others.

Let (X,x) be a reduced germ with good $\mathbb{C}^*$-action of dim n. Denote by $\xi \in \theta_{X,x}$ the infinitesimal generator of the $\mathbb{C}^*$-action: ξ is characterized by the property that it is multiplication by k on R_k. This is also called the *euler derivation*. The following lemma is due to Naruki (1977).

(9.9) *Lemma.* Put $\tilde{\Omega}^p_{X,x} := \Omega^p_{X,x}/\text{torsion}$. Then the complexes

$$0 \to \mathbb{C} \to \mathcal{O}_{X,x} \xrightarrow{d} \tilde{\Omega}^1_{X,x} \xrightarrow{d} \cdots \xrightarrow{d} \tilde{\Omega}^n_{X,x} \to 0 \quad \text{and}$$

$$0 \to \tilde{\Omega}^n_{X,x} \xrightarrow{\iota_\xi} \tilde{\Omega}^{n-1}_{X,x} \xrightarrow{\iota_\xi} \cdots \xrightarrow{\iota_\xi} \mathcal{O}_{X,x} \xrightarrow{\alpha} \mathbb{C} \to 0,$$

are exact, where α is 'take the degree 0 part', more precisely,

$$\alpha(\phi)(z) = \int_0^1 (e^{2\pi i\theta}.\phi)(z)d\theta.$$

Proof. The $\mathbb{C}^*$-action on $\mathcal{O}_{X,x}$ extends in an obvious way to the complex $\tilde{\Omega}^\cdot_{X,x}$. This action has positive weights (except on the constants in $\mathcal{O}_{X,x}$) and the Lie derivative L_ξ acts on the weight w subspace of $\tilde{\Omega}^p_{X,x}$ as multiplication by w. So if we define $h : \tilde{\Omega}^\cdot_{X,x} \to \tilde{\Omega}^{\cdot -1}_{X,x}$ by letting h be on the weight w subspace $\frac{1}{w}\iota_\xi$ $(w \neq 0)$ resp. 0. $(w = 0)$, then we see that h defines a homotopy equivalence between $\tilde{\Omega}^\cdot_X$ and the subcomplex of $\tilde{\Omega}^\cdot_{X,x}$ given by the constants (in degree 0). This proves the exactness of the first complex. The exactness of the second complex is proved in the same way.

Now let (X_0,x) be an icis of dim $n \geq 2$ and let f denote a miniversal deformation of (X_0,x). According to (9.6) there is a natural $\mathcal{O}_{X_0,x}$-linear pairing

$$< , > : \theta_{X_0,x}/(\theta_{f,x}\otimes\mathcal{O}_{X_0,x}) \times \omega_{X_0,x}/\Omega^n_{X_0,x} \to H^1_{\{x\}}(\Omega^{n-1}_{X_0,x})$$

induced by the contraction mapping. Moreover, (9.6) asserts that if ω_0 is a generator of $\omega_{X_0,x}$ with image ω_0' in $\omega_{X_0,x}/\Omega^n_{X_0,x}$, then $\xi' \in \theta_{X_0,x}/(\theta_{f,x}\otimes\mathcal{O}_{X_0,x}) \to <\xi',\omega_0'>$ is an isomorphism.

(9.10) *Proposition.* Suppose that in the above situation, (X_0,x) has a good $\mathbb{C}^*$-action with infinitesimal generator ξ_0. Then the image ξ_0' of ξ_0

in $\theta_{X_0,x}/(\theta_{f,x}\otimes\mathcal{O}_{X_0,x})$ is an $\mathcal{O}_{X_0,x}$-generator and $\omega' \in H^1_{\{x\}}(\Omega^n_{X_0,x}) \to \langle\xi_0',\omega'\rangle$ is an isomorphism so that $\langle\ ,\ \rangle$ is perfect. Furthermore, $\mu(X_0,x) = \tau'(X_0,x) = \tau(X_0,x)$.

Proof. We first show that $\omega' \to \langle\xi_0',\omega'\rangle$ is injective. If $\omega \in \omega_{X_0,x}$ is such that $\iota_{\xi_0}(\omega) \in \tilde{\Omega}^{n-1}_{X_0,x}$, then $\iota_{\xi_0}(\iota_{\xi_0}(\omega)) = 0$ and so by the preceding lemma, $\iota_{\xi_0}(\omega) = \iota_{\xi_0}(\omega_1)$ for some $\omega_1 \in \tilde{\Omega}^n_{X_0,x}$. Hence $\iota_{\xi_0}(\omega-\omega_1) = 0$. Since ξ_0 is nowhere zero on a punctured neighbourhood of x in X_0, it follows that $\omega-\omega_1$ vanishes on a punctured neighbourhood, which just means that $\omega = \omega_1 \in \tilde{\Omega}^n_{X_0,x}$.

Next we prove that $\omega' \to \langle\xi_0',\omega'\rangle$ is surjective. Let $\alpha \in j_*j^{-1}\Omega^{n-1}_{X_0,x}$ (where $j: X_0-\{x\} \subset X_0$). Then $\iota_{\xi_0}(\alpha) \in j_*j^{-1}\Omega^{n-2}_{X_0,x} = \tilde{\Omega}^{n-2}_{X_0,x}$ (by (8.17)). Since $\iota_{\xi_0}(\iota_{\xi_0}(\alpha)) = 0$, there is by (9.9) an $\alpha_1 \in \tilde{\Omega}^{n-1}_{X_0,x}$ with $\iota_{\xi_0}(\alpha) = \iota_{\xi_0}(\alpha_1)$. Then there is a unique $\omega \in j_*j^{-1}\tilde{\Omega}^n_{X_0,x} = \omega_{X_0,x}$ with $\iota_{\xi_0}(\omega) = \alpha-\alpha_1 \equiv \alpha \pmod{\tilde{\Omega}^{n-1}_{X_0,x}}$ (see (9.1)).

As we know already that $\xi' \to \langle\xi',\omega_0'\rangle$ is an isomorphism, it follows that $\langle\ ,\ \rangle$ is perfect. For any ξ' we find a unique ω' such that $\langle\xi',\omega_0'\rangle = \langle\xi_0',\omega'\rangle$. If we write $\omega' = \phi\omega_0'$ with $\phi \in \mathcal{O}_{X_0,x}$ then it follows that $\xi' = \phi\xi_0'$ so that ξ_0' generates. It also follows that the three modules in question have the same $\mathbb{C}$-dimension: so $\tau(X_0,x) = \tau'(X_0,x)$. By lemma (9.9), $\tilde{\Omega}^n_{X_0,x} = d\tilde{\Omega}^{n-1}_{X_0,x}$ and hence $H^1_{\{x\}}(\Omega^n_{X_0,x}) = H^1_{\{x\}}(d\Omega^{n-1}_{X_0,x})$, showing that $\tau'(X_0,x) = \mu(X_0,x)$ also.

For $n = 1$, we can still show that ξ_0' generates $\theta_{X_0,x}/(\theta_f\otimes\mathcal{O}_{X_0,x})$ and that $\langle\ ,\ \rangle$ has the property that $\omega' \to \langle\xi_0',\omega'\rangle$ is injective. From this one concludes in the same way as above that $\tau(X_0,x)$, $\tau'(X_0,x)$ and $\mu(X_0,x)$ all coincide (we omit the proofs).

9.C *A period mapping*

In this section we define a so-called period mapping with the help of which we shall find an alternative description of the discriminant of a Kleinian singularity.

Let $f : X \to S$ be a good Stein representative of a miniversal deformation of an icis (X_0,x) of dim $n \geq 1$ and let $s_0 \in S-D$. The monodromy representation $\rho : \pi(S-D,s_0) \to \mathrm{Aut}\, H_n(X_{s_0})$ determines a regular covering $\pi : \widetilde{S-D} \to S-D$: any $\tilde{s} \in \widetilde{S-D}$ over $s \in S-D$ is represented by a continuous path α in S-D from s_0 to s and any two such, α,α', represent the same $\tilde{s}$ if and only if there is no monodromy along the loop $\alpha * \alpha'^{-1}$ (i.e $[\alpha * \alpha'^{-1}] \in \mathrm{Ker}(\rho)$). The covering group of π is precisely the image Γ of ρ. Now suppose that we are also given a section ω of ω_f over X. Then for each $s \in S$ we find a De Rham cohomology class $[\omega|X_{s,reg}] \in H^n(X_{s,reg};\mathbb{C})$. If $\tilde{s} \in \widetilde{S-D}$ is represented by a path α, then $[\omega|X_s]$ can be transported along α to give a well-defined element of $H^n(X_{s_0};\mathbb{C})$. Thus we find a map

$$\tilde{P}_\omega : \widetilde{S-D} \to H^n(X_{s_0};\mathbb{C}).$$

An equivalent way to describe $\tilde{P}_\omega(\tilde{s})$ is via the identification of $H^n(X_{s_0};\mathbb{C})$ with $\mathrm{Hom}(H_n(X_{s_0}),\mathbb{C})$: if $\gamma \in H_n(X_{s_0})$, then

$$\tilde{P}_\omega(\tilde{s})(\gamma) = \int_{\alpha_*(\gamma)} \omega|X_s$$

where $\alpha_*(\gamma)$ denotes the class in $H_n(X_s)$ obtained by displacing γ along α. We call $\tilde{P}_\omega$ the *period map* defined by ω. As $\tilde{P}_\omega$ is just 'viewing ω as a section of $R^n f_*(\mathbb{C}_X) \otimes_{\mathbb{C}} \mathcal{O}_{S-D}$', $\tilde{P}_\omega$ is holomorphic.

(9.11) *Proposition.* Suppose that $H^1_{\{x\}}(d\Omega^{n-2}_{X_0,x}) = 0$ (a condition satisfied if $H^n(X_0,X_0-\{x\};\mathbb{C}) = 0$) and that $\omega \in \Gamma(X,\omega_f)$ generates $\omega_{f,x}$ over $\mathcal{O}_{X,x}$. Then there exists a neighbourhood V of 0 in S such that $\tilde{P}_\omega|\pi^{-1}(V\cap(S-D))$ is a local isomorphism.

Proof. Let $\eta_1,\ldots,\eta_\tau$ be an $\mathcal{O}_{S,0}$-basis of $\Theta_{S,0}\langle D\rangle$ (with $\tau = \tau(X_0,x) = \dim S$). Then (9.7) implies that $\nabla_{\eta_1}[\omega],\ldots,\nabla_{\eta_\tau}[\omega]$ is part of an $\mathcal{O}_{S,0}$-basis of the free $\mathcal{O}_{S,0}$-module $\omega_{f,x}/d\Omega^{n-1}_{f,x}$. By the same argument as used in (8.6), $\omega_{f,x}/d\Omega^{n-1}_{f,x}$ is just the stalk of the coherent $\mathcal{O}_S$-module $f_*\omega_f/d(f_*\Omega^{n-1}_f)$ in $0 \in S$. So there exists a neighbourhood V of 0 in S such that $\eta_1,\ldots,\eta_\tau$ converge on V and $\nabla_{\eta_1}[\omega],\ldots,\nabla_{\eta_\tau}[\omega]$ generate a direct summand of rank τ in $f_*\omega_f/d(f_*\Omega^{n-1}_f)$. This implies that the partial derivatives of $\tilde{P}_\omega$ with respect to $\eta_1,\ldots,\eta_\tau$ are linearly independent everywhere in $\pi^{-1}(V\cap(S-D))$ so $\tilde{P}_\omega$ is there an immersion.

In certain cases we shall want to know that $\tilde{P}_\omega$ extends to a larger space. In this respect, the following lemma will prove useful.

(9.12) *Lemma*. Let $f : (\mathbb{C}^{n+k},0) \to (\mathbb{C}^k,0)$ be defined by $(z_1,\ldots,z_{n+k}) \to (z_1^2+\ldots+z_{n+1}^2, z_{n+2},\ldots,z_{n+k})$ with $n > 0$ even and let $\omega \in \omega_{f,0}$ be a generator. Then there is a good representative $f : X \to S$ of f with ω converging on X such that if $\delta(s) \in H_n(X_s)$, $s \in S-D$, denotes a generator, the function $s \in S-D \to (\int_{\delta(s)}\omega|X_s)^2$ extends holomorphically over S and has divisor equal to nD.

Proof. Let $\omega_0 \in \omega_f$ be defined by $df_1 \wedge \omega_0 = dz_1 \wedge\ldots\wedge dz_{n+1}$. Then ω_0 generates $\omega_{f,0}$ and the image of ω_0 in $\omega_{f,0}/d\Omega^{n-1}_{f,0}$ generates the latter over $\mathcal{O}_{\mathbb{C}^k,0}$. So there exists a $\phi \in \mathcal{O}_{\mathbb{C}^k,0}$ such that $\omega-\phi\omega_0 \in d\Omega^{n-1}_{f,0}$. By restricting this to $(X_0,0)$ we see that we must have $\phi(0) \neq 0$. So it suffices to prove the lemma for ω_0. As the function in question then only depends on the first coordinate, we may as well assume that $k = 1$.

If $f : X \to S$ is a good representative, then the calculation, done in (8.26) (give $z_1,\dots,z_{n+1}$ weight 1 and f_1 weight 2; $z_{n+2},\dots,z_{n+k}$ are 'dummy' variables) shows that upon writing $s = (t_1,\dots,t_k)$ we have

$$\int_{\delta(s)} \omega_0 | X_s = ct_1^{\frac{1}{2}n},$$

for some $c \in \mathbf{C}^*$. This proves the lemma.

We now concentrate on the case of Kleinian singularity. We begin by noting that the equations given in 1, ex. 2.3 are weighted homogeneous relative the following degrees

type	$f(z_1,z_2,z_3)$	weights z_1,z_2,z_3	degree f
A_ℓ	$z_1^{\ell+1} + z_2z_3$	$1,\ 1,\ \ell$	$\ell+1$
D_ℓ	$z_1^{\ell-1} + z_1z_2^2 + z_3^2$	$2,\ \ell-2,\ \ell-1$	$2\ell-2$
E_6	$z_1^4 + z_2^3 + z_3^2$	3, 4, 6	12
E_7	$z_1^3z_2 + z_2^3 + z_3^2$	4, 6, 9	18
E_8	$z_1^5 + z_2^3 + z_3^2$	6, 10, 15	30

If ℓ is the Milnor number of $(f^{-1}(0),0)$, then choose monomials $\phi_2,\dots,\phi_\ell \in \mathbf{C}[z_1,z_2,z_3]$ such that $1,\phi_2,\dots,\phi_\ell$ map to a $\mathbf{C}$-basis of $\mathbf{C}\{z_1,z_2,z_3\}/(\frac{\partial f}{\partial z_1}, \frac{\partial f}{\partial z_2}, \frac{\partial f}{\partial z_3})\mathbf{C}\{z_1,z_2,z_3\}$ and use these to extend f to a miniversal deformation $F : X := \mathbf{C}^{\ell+2} \to \mathbf{C}^\ell =: S$ by $(z_1,z_2,z_3,u_2,\dots,u_\ell) \to (f(z) + u_2\phi_2(z) + \dots + u_\ell\phi_\ell(z), u_2,\dots,u_\ell)$. There are obvious $\mathbf{C}^*$-actions on X and S making F equivariant: give u_λ and t_λ $(\lambda \geq 2)$ weight $\deg(f)-\deg(\phi_\lambda)$ and give t_1 weight $\deg(f)$. One checks that the weights of $t_1,\dots,t_\ell$ are relatively prime so that the $\mathbf{C}^*$-action is effective. Moreover, all weights are > 0 and so both $\mathbf{C}^*$-actions are good. We let ω be the generating section of ω_f characterized by $dF_1 \wedge \omega = dz_1 \wedge dz_2 \wedge dz_3$ so that ω has weight $\deg(z_1z_2z_3)-\deg(f)$. We observe from the table above that in all cases this is equal to 1. We take $s_0 := (1,0,\ldots,0) \in S-D$, and we let $\tilde{P} : \widetilde{S-D} \to H^2(X_{s_0};\mathbb{C})$ be the period map defined by ω. Since the monodromy group $\Gamma \subset \mathrm{Aut}\, H_2(X_{s_0})$ is finite, the orbit space $\Gamma\backslash H^2(X_{s_0};\mathbb{C})$ exists as an affine algebraic variety. Scalar multiplication in $H^2(X_{s_0},\mathbb{C})$ endows the latter naturally with a $\mathbb{C}^*$-action. Clearly $\tilde{P}$ drops to a map $P : S-D \to \Gamma\backslash H^2(X_{s_0};\mathbb{C})$. Since ω has weight 1, P is $\mathbb{C}^*$-equivariant. Before we continue our study of P we first have to recall some of the invariant theory of finite reflection groups. Our general reference is Bourbaki (Lie V).

(9.13) Let V be a real vector space of finite dimension ℓ and let W be a finite group of automorphisms generated by reflections. We suppose W irreducible: there is no proper subspace $\neq 0$ of V left invariant by W. We let W act in the obvious way on the symmetric algebra $S_*(V) = \oplus_{k=0}^{\infty} S_k(V)$ of V and in the contragredient manner on the dual V* of V.

Let $S_*(V)^W$ resp. $S_*(V)^{-W}$ denote the set of $\phi \in S_*(V)$ satisfying $w.\phi = \phi$ resp. $w.\phi = \det(w)\phi$ for all $w \in W$. It is clear that $S_*(V)^W$ is a subalgebra of $S_*(V)$ and that $S_*(V)^{-W}$ is an $S_*(V)^W$-module. An interesting element $d \in S_*(V)^{-W}$ is obtained by choosing for any reflection $s \in W$ a (-1)-eigenvector δ_s and by letting d be the product of these (Lie V, §5, no. 4). The degree of d is equal to the number of reflections in W, which is given by $\frac{1}{2}h\ell$, where h is the so-called Coxeter number of W. Then according to Chevalley (1955), $S_*(V)^W$ is a polynomial algebra, $S_*(V)$ is a free $S_*(V)^W$-module of rank $|W|$ and the discriminant ideal of the extension $S_*(V)^W \subset S_*(V)$ is generated by $d^2 \in S_{h\ell}(V)^W$.

In geometric terms this means that the orbit space $W\backslash V^*_{\mathbb{C}}$ is an affine algebraic variety isomorphic to $\mathbb{C}^\ell$ and if $D_W \subset W\backslash V^*_{\mathbb{C}}$ denotes the discriminant of the orbit map $\pi : V^*_{\mathbb{C}} \to W\backslash V^*_{\mathbb{C}}$, then $\pi^{-1}(D_W) = \Sigma_{s\in W} 2H_s$, (equality of divisors), where the sum is over all the reflections s in W and $H_s \subset V^*_{\mathbb{C}}$ denotes the hyperplane left pointwise fixed by s.

(9.14) *Theorem*. The map $P : S-D \to \Gamma\backslash H^2(X_{s_0};\mathbb{C})$ extends to a $\mathbb{C}^*$-equivariant isomorphism from S to $\Gamma\backslash H^2(X_{s_0};\mathbb{C})$ mapping D onto D_Γ.

The proof is divided into four steps.

Step 1. $\tilde{P}$ is a local isomorphism.

Proof. Let $s \in S-D$ and consider the exact sequence

$$\begin{array}{ccccccccc} 0 \to & H_2(\partial\overline{X}_s) & \to & H_2(\overline{X}_s) & \xrightarrow{j} & H_2(\overline{X}_s,\partial\overline{X}_s) & \to & H_1(\partial\overline{X}_s) & \to 0 \\ & \wr\Vert & & & & \wr\Vert & & & \\ & H_2(\partial\overline{X}_0) & & & & H_2(\overline{X}_s)^* & & & \end{array}$$

The intersection form on $\overline{X}_s$ is negative definite, so its adjoint j is injective. This implies that $H^2(X_0,X_0-\{x_0\}) \cong H^1(\partial\overline{X}_0) \cong H_2(\partial\overline{X}_0) = 0$. (This can also be derived from the fact that X_0 is a quotient singularity.) Then prop. (9.11) implies that $\tilde{P}$ is a local isomorphism over a neighbourhood of 0 in S. Because of the $\mathbb{C}^*$-actions around, it follows that $\tilde{P}$ is a local isomorphism on all of $\widetilde{S-D}$.

Step 2. $P : S-D \to \Gamma\backslash H^2(X_{s_0};\mathbb{C})$ extends over S and maps D into D_Γ.

Proof. It suffices to check this at a regular point s_1 of D. Then X_{s_1} contains precisely one singular point y and this point is quadratic. Let $f : \mathcal{Y} \to V$ be a good representative of the germ of f at y so that (9.12) applies to it. Let $s \in V \cap (S-D)$ and consider the Mayer-Vietoris sequence

$$\ldots \to H_2(X_s-Y_s) \oplus H_2(Y_s) \to H_2(X_s) \to H_2(\partial Y_s) \to \ldots$$

Since $H_2(\partial Y_s;\mathbb{Q}) = 0$, it follows that $H_2(X_s;\mathbb{Q})$ is spanned over $\mathbb{Q}$ by the images of $H_2(X_s - Y_s)$ and $H_2(Y_s)$. If $\gamma \in H_2(X_s - Y_s)$, then clearly γ extends to a continuous family $\{\gamma(s') \in H_2(X_{s'} - Y_{s'}) : s' \in V\}$ with $\gamma = \gamma(s)$ so that $s' \in V \to \int_{\gamma(s')} \omega | X_{s'}$ is holomorphic on V. If $\delta(s)$ is one of the (two) generators of $H_2(Y_s)$, then (9.12) asserts that $s \in V \to \int_{\delta(s)} \omega | Y_s$ is holomorphic on V and vanishes on $V \cap D$. This proves that P extends holomorphically over S and that $P(D) \subset D_\Gamma$.

Step 3. The divisors D and D_Γ have the same weighted degree (= degree of a weighted homogeneous defining equation) and $P^*(D_\Gamma) = D$.

Proof. If $\delta \in \mathbb{C}[t_1, \ldots, t_\ell]$ is a homogeneous defining equation for D then $\delta(t_1, 0, \ldots, 0) = \text{const}.t_1^\ell$ for D has multiplicity ℓ in 0. This implies that $\deg(\delta) = \ell.\deg(t_1)$. We look at the tables in Bourbaki (Lie VI) and find that in each case $\deg(t_1) = \deg(f)$ is just the Coxeter number h of Γ. So $\deg(\delta) = \ell h = \deg(D_\Gamma)$.

Since $\widetilde{P}$ is a local isomorphism, P will not have its image in D_Γ so that $P^*(D_\Gamma)$ is a divisor on S. Since P^* is $\mathbb{C}^*$-equivariant, $P^*(D_\Gamma)$ has the same weighted degree as D_Γ. In step 2 we found that $P^*(D_\Gamma)$ is of the form D+D' where D' is an effective divisor. As $P^*(D_\Gamma)$ and D have the same degree, we must have $\deg(D') = 0$ and so $D' = 0$.

Step 4. Proof of the theorem.

It follows from steps 1 and 3 that $P|S-D$ is a local isomorphism. So the ramification divisor R of P is supported by D. If $R \neq 0$, then R would occur with multiplicity > 1 in $P^*(D_\Gamma)$ which is impossible, since $P^*(D_\Gamma) = D$ is reduced. Hence P is a local isomorphism everywhere on S. In particular P maps a neighbourhood of $0 \in S$ isomorphically onto a neighbourhood of $\Gamma.0$ in $\Gamma \backslash H^2(X_{s_0};\mathbb{C})$. Since P is $\mathbb{C}^*$-equivariant, it follows that P is a global isomorphism.

(9.15) The preceding theorem is one of the major implications of a conjecture made by Grothendieck around 1969. This conjecture was phrased in the context of linear algebraic groups, so we shall not state it here. A proof of Grothendieck's conjecture was announced by Brieskorn (1970b). Slodowy (1980) gave a complete account of a proof of Grothendieck's conjecture based on Brieskorn's ideas. The proof of (9.14) appears in Looijenga (1974). The identification between (S,D) and $(\Gamma\backslash H_2(X_{s_0};\mathbb{C}),D_\Gamma)$ is of interest for various reasons. For instance, Deligne (1972) has shown that the space $V^*_{\mathbb{C}}-\cup H_s$ (the situation and notation are those of (9.13)) has a contractible universal covering space. This implies that the same is true for $\Gamma\backslash H_2(X_{s_0};\mathbb{C})-D_\Gamma \cong S-D$. Therefore the fundamental group of S-D is a complete homotopy invariant of S-D. These groups are quite interesting, and have been the subject of further investigation: Brieskorn (1971), Brieskorn & Saito (1972), Deligne (1972). The result of Deligne supports a conjecture saying that the complement of the discriminant of a miniversal deformation of a positive-dimensional icis has always a contractible covering. The status of this conjecture remains somewhat in the doubt because there are so few cases in which it has been verified. Knörrer (1982a) has shown that the conjecture is not true for the simplest zero-dimensional icis which is not a hypersurface: $(z_1z_2) \to (z_1^2,z_2^2)$. Theorem (9.14) is also the point of departure for proving similar results for icis's of dim 2 which are higher up in the hierarchy (simply-elliptic singularities, cusps, triangle singularities,...), see Looijenga (1978, 1981, 1983). If we go beyond a certain range, then paying attention to the behaviour of just one element of ω_f may not be a wise thing to do; it is probably better to replace a generator of ω_f by a subspace of $\omega_{f,x}/d\Omega^{n-1}_{f,x}$ of dim $\frac{1}{2}(\mu_0+\mu_+)$ (the notation is that of the discussion (7.12); it can be shown that $\mu_0+\mu_+$ is always even). If we increase n instead, then picking a generator of ω_f is not a good choice either. In the even-dimensional hypersurface case where the semi-universal deformation is of the form $(z_1,\ldots,z_{2m+1},u_2,\ldots,u_\tau) \to (g(z,u),u_2,\ldots,u_\tau)$, Saito has proposed to consider instead $(\nabla_{\partial/\partial t_1})^{m-1}\omega$, $\omega \in \omega_{f,0}$ a generator. This has been subsumed in his theory of primitive forms (Saito, 1982).

REFERENCES

A'Campo, N. (1973): Sur la monodromie des singularités isolées d'hypersurfaces complexes. Invent. Math. 20, 147 - 169.

A'Campo, N. (1975): La fonction zêta d'une monodromie. Comment. Math. Helv. 50, 233 - 248.

Arnol'd, V.I. (1974): Critical points of smooth functions. In Proc. Intern. Congr. Math. Vancouver, 19 - 39.

Arnol'd, V.I. (1976): Local normal forms of functions. Invent. Math. 35, 87 - 109.

Bourbaki, N.: Algèbre, Ch. X. Paris: Masson.

Bourbaki, N.: Algèbre Commutative, Ch. V, VI. Paris: Hermann.

Bourbaki, N.: Groupes et Algèbres de Lie, Ch. IV, V et VI. Paris: Masson.

Brieskorn, E.V., (1970a): Die Monodromie der isolierten Singularitäten von Hyperflächen. Manuscripta Math. 2, 103 - 161.

Brieskorn, E.V., (1970b): Singular elements of semi-simple algebraic groups. In Proc. Intern. Congress Math. Nice, Vol. 2, 279 - 284. Paris: Gauthier-Villars.

Brieskorn, E.V., (1971): Die Fundamentalgruppe des Raumes der regulären Orbits einer endlichen komplexen Spiegelungsgruppe. Invent. Math. 12, 57 - 61.

Brieskorn, E.V. & Saito, K., (1972): Artin-Gruppen und Coxeter-Gruppen. Invent. Math. 17, 245 - 271.

Brieskorn, E.V., (1979): Die Hierarchie der 1-modularen Singularitäten. Manuscripta Mathematica 27, 183 - 219.

Buchsbaum, D.A. & Rim, D.S., (1964): A generalized Koszul complex II. Trans. Am. Math. Soc. 111, 197 - 224.

Buchweitz, R.-O. & Greuel, G.-M., (1980): The Milnor number and deformations of complex curve singularities. Invent. Math. 58, 241 - 281.

Cartan, H., (1957): Quotient d'un espace analytique par un groupe d'automorphismes. In Algebraic Geometry and Topology, ed. R.H. Fox et alii, pp. 90 - 102. Princeton, N.J.: Princeton U.P.

Chevalley, C., (1955): Invariants of finite groups generated by reflections. Am. J. of Math. 77, 778 - 782.

Chmutov, S.V., (1982): Monodromy groups of critical points of functions. Invent. Math. 67, 123 - 131.

Clemens, H., (1969): Picard-Lefschetz theorem for families of non-singular algebraic varieties acquiring ordinary singularities. Trans. Am. Math. Soc. 136, 93 - 108.

Deligne, P., (1970): Equations Différentielles à Points Réguliers Singuliers. Lecture Notes in Math. 163. Berlin etc.: Springer Verlag.

Deligne, P., (1972): Les immeubles des groupes de tresses généralisés. Invent. Math. 17, 273 - 302.

Dolgachev, I.V., (1974): Conic quotient singularities of complex surfaces. Funk. Anal. 8, no. 2, 75 - 76.

Dolgachev, I.V., (1975): Automorphic forms and quasi-homogeneous singularities. Funk. Anal. 9, no. 2, 67 - 68.

Durfee, A., (1979): Fifteen characterizations of rational double points and simple critical points. Ens. Math. 25, 131 - 163.

Du Val, P., (1934): On isolated singularities of surfaces which do not affect the conditions of adjunction. Proc. Camb. Phil. Soc. 30, 453 - 459.

Ebeling, W., (1981): Quadratische Formen und Monodromiegruppen von Singularitäten. Math. Ann. 255, 463 - 498.

Ebeling, W., (1983): Arithmetic Monodromy Groups, to appear.

Essen, A. van den, (1979): Fuchsian Modules. Thesis, University of Nijmegen.

Forster, O. & Knorr, K., (1971): Ein Beweis des Grauertschen Bildgarbensatzes nach Ideen von B. Malgrange. Manuscripta Math. 5, 19 - 44.

Fox, R.H., (1952): On Fenchel's conjecture about F-groups. Mat. Tidsskrift B, 61 - 65.

Gabrielov, A.M., (1974): Dynkin diagrams of unimodal singularities. Funk. Anal. i Pril. 8, no. 3, 1 - 6.

Gibson et alii (1976): Topological Stability of Smooth Mappings. Lecture Notes in Mathematics 552. Berlin etc.: Springer Verlag.

Giusti, M., (1977): Classification des singularités isolées d'intersections complètes simples. C.R. Acad. Sc. Paris 284, 167 - 169.

Godement, R., (1958): Topologie Algébrique et Théorie des faisceaux. Paris: Hermann.

Gonzalez-Sprinberg, G. & Verdier, J.-L., (1981): Points doubles rationnels et représentations de groupes. C.R. Acad. Sc. Paris 293, 111 - 113.

Grauert, H. & Remmert, R., (1971): Analytische Stellenalgebren. Grundlehren 176. Berlin etc.: Springer Verlag.

Greuel, G.-M., (1975): Der Gauss-Manin-Zusammenhang isolierter Singularitäten von vollständigen Durchschnitten. Math. Ann. 214, 235 - 266.

Greuel, G.-M., (1980): Dualität in der lokalen Kohomologie isolierter Singularitäten. Math. Ann. 250, 157 - 173.

Greuel, G.-M. & Steenbrink, J., (1983): On the topology of smoothable singularities. In Proc.Symp.Pure Math. Vol.XL. Singularities at Arcata, ed. P. Orlik, part 1, 535-545: Providence, R.I.: AMS.

Gromoll, D. & Meyer, W., (1969): On differentiable functions with isolated critical points. Topology 8, 361 - 370

Grothendieck, A., (1967): Local Cohomology. Lecture Notes in Math. 41. Berlin etc.: Springer Verlag.

Gunning, R.C., (1974): Lectures on Complex Analytic Varieties: Finite Analytic Mappings. Princeton, N.J.: Princeton U.P.

Hamm, H., (1969): Die Topologie isolierter Singularitäten von vollständigen Durchschnitten komplexer Hyperflächen. Thesis, Bonn.

Hamm, H., (1971a): Topology of isolated singularities of complex spaces. In Proc. Liverpool Singularities Symposium, vol. II, ed. C.T.C. Wall, pp. 213 - 217. Lecture Notes in Mathematics, 209. Berlin, Heidelberg, New York: Springer Verlag.

Hamm, H., (1971b): Lokale topologische Eigenschaften komplexer Räume. Math. Ann. 191, 235 - 252.

Hamm, H., (1974): Habilitationsschrift, Göttingen.

Herrera, M., (1966): Integration on a semi-analytic set. Bull. Soc. Math. de France 94, 141 - 180.

Hironaka, H., (1974): Triangulation of algebraic sets. In Proc. Symp. Pure Math. Vol. XXIX. Algebraic Geometry, Arcata. Providence, R.I.: Am. Math. Soc.

Husemoller, D., (1966): Fibre Bundles. New York, N.Y. etc.: McGraw-Hill.

Janssen, W., (1983): Skew symmetric vanishing lattices and their monodromy groups, to appear.

Karras, U., (1977): Deformations of cusp singularities. In Proc. Symp. Pure Math. XXX, ed. R.O. Wells jr. pp. 37 - 40. Providence, R.I.: Am. Math. Soc.

Kiehl, R. & Verdier, J.-L., (1971): Ein einfacher Beweis des Kohärenzsatzes von Grauert. Math. Ann. 195, 24 - 50.

Klein, F., (1884, 1956): Lectures on the Isocahedron and the Solution of Equations of the Fifth Degree (2nd edition). New York N.Y.: Dover Publ. Inc.

Knörrer, H., (1982a): Zum K(π,1)-Problem für isolierte Singularitäten von vollständigen Durchschnitten. Comp. Math. 45, 333 - 340.

Knörrer, H., (1982b): Group representations and resolution of rational double points. Preprint, Bonn.

Lamotke, K., (1975): Die Homologie isolierter Singularitäten. Math. Z. 143, 27 - 44.

Landman, A., (1973): On Picard-Lefschetz transformation for algebraic manifolds acquiring general singularities. Trans. Am. Math. Soc. 181, 89 - 126.

Lê, D.T., (1973): Calcul du nombre de cycles évanouissants d'une hypersurface complexe. Ann. Inst. Fourier 23, no. 4, 261 - 270.

Lê, D.T., (1978): The geometry of the monodromy theorem. In C.P. Ramanujam, a tribute, ed. K.G. Ramanathan. Tata Institute Studies in Math. 8, Berlin etc.: Springer Verlag.

Lê, D.T., (1981): Faisceaux constructibles quasi-unipotents. Sém. Bourbaki, exposé 581.

Lefschetz, S., (1924): L'Analysis Situs et la Géométrie Algébrique. Paris: Gauthier-Villars.

Looijenga, E., (1974): A period mapping for certain semi-universal deformations. Comp. Math. 30, 299 - 316.

Looijenga, E., (1977): On the semi-universal deformation of a simple elliptic singularity I: Unimodularity. Topology 16, 257 - 262.

Looijenga, E., (1978): On the semi-universal deformation of a simple elliptic singularity II: The discriminant. Topology 17, 23 - 40.

Looijenga, E., (1981): Rational surfaces with an anti-canonical cycle. Ann. of Math. 114, 267 - 322.

Looijenga, E., (1983): The smoothing components of a triangle singularity, I. In Proc. Symp. Pure Math. Vol. XL: Singularities at Arcata. ed. P. Orlik, part 2, 173 - 183. Providence, R.I.: AMS.

Malgrange, B., (1973): Letter to the editors. Invent. Math. 20, 171 - 172.

Malgrange, B., (1974a): Sur les points singuliers des équations différentielles. Ens. Math. 20, 147 - 176.

Malgrange, B., (1974b): Intégrales asymptotiques et monodromie. Ann. Sc. Ec. Norm. Sup. 7, 405 - 430.

Mather, J., (1968): Stability of C^∞-mappings III. Publ. Math. de l'I.H.E.S. 35, 127 - 156.

Mather, J., (1969): Stability of C^∞-mappings IV. Publ. Math. de l'I.H.E.S. 37, 223 - 248.

Mather, J., (1970): Stability of C^∞-mappings V. Adv. in Math. 4, 301 - 331.

Mather, J., (1971): Stability of C^∞-mappings VI. In Proc. Liverpool Singularities Symposium, ed. C.T.C. Wall, pp. 207 - 253. Lecture Notes in Math. 192. Berlin etc.: Springer Verlag.

Mather, J., (1973): Stratifications and mappings. In Dynamical Systems, pp. 195 - 223. Ed. M.M. Peixoto. New York, N.Y.: Academic Press.

Matsumura, H., (1980): Commutative Algebra (2nd edition). Reading, Mass.: Benjamin Cummings.

Mérindol, J.-Y., (1982): Les singularités simples elliptiques, leurs déformations, les surfaces de Del Pezzo et les transformations quadratiques. Ann. Sc. Ec. Norm. Sup. 15, 17 - 44.

Milnor, J.W., (1963): Morse Theory. Ann. of Math. Studies 51. Princeton N.J.: Princeton U.P.

Milnor, J.W., (1968): Singular Points of Complex Hypersurfaces. Ann. of Math. Studies 61. Princeton, N.J.: Princeton U.P.

Milnor, J.W., (1975): On the 3-dimensional Brieskorn manifolds M(p,q,r). In Knots, Groups and 3-manifolds ed. L.P. Neuwirth, pp. 175 - 225. Princeton, N.J.: Princeton U.P.

Mumford, D., (1976): Algebraic Geometry I: Complex Projective Varieties. Grundlehren 221, Berlin, Heidelberg, New York: Springer Verlag.

Munkres, J.R., (1966): Elementary Differential Topology (rev. edn.). Ann. of Math. Studies 54. Princeton N.J.: Princeton U.P.

Narasimhan, R., (1966): Introduction to the Theory of Analytic Spaces. Lecture Notes in Math. 25. Berlin etc.: Springer Verlag.

Naruki, I., (1977): Some remarks on isolated singularity and their application to algebraic manifolds. Publ. Res. Inst. Math. Sc. 13, no. 3, 17 - 46.

Neumann, W., (1983): Geometry of quasi-homogeneous surface singularities. In Proc. Symp. Pure Math. Vol. XL: Singularities at Arcata. ed. P. Orlik, part 2, 245 - 258. Providence, R.I.: AMS.

Orlik, P. & Wagreich, P., (1971): Isolated singularities of algebraic surfaces with $\mathbb{C}^*$-action. Ann. of Math. 93, 205 - 228.

Pham, F., (1979): Singularités des Systèmes Différentiels de Gauss-Manin. Progress in Math. 2. Boston: Birkhäuser.

Picard, E. & Simart, S., (1897): Traité des Fonctions Algébriques de Deux Variables. Vol. I. Paris: Gauthier-Villars.

Pinkham, H.C., (1974): Deformations of algebraic varieties with $\mathbb{G}_m$ action. Astérisque 20. Soc. Math. de France.

Pinkham, H.C., (1977a): Normal surface singularities with $\mathbb{C}^*$-action. Math. Ann. 227, 183 - 193.

Pinkham, H.C., (1977b): Singularités exceptionnelles, la dualité étrange d'Arnold et les surfaces K-3. C.R. Acad. Sc. Paris 284, 615 - 618.

Pinkham, H.C., (1977c): Groupe de monodromie des singularités unimodulaires exceptionnelles. C.R. Acad. Sc. Paris 284, 1515 - 1518.

Pinkham, H.C., (1977d): Simple elliptic singularities, Del Pezzo surfaces and Cremona transformations. In Proc. Symp. Pure Math. XXX ed. R.O. Wells jr., pp. 69 - 71. Providence, R.I.: Am. Math. Soc.

Rim, D.S., (1972): Torsion differentials and deformations. Trans. Am. Math. Soc. 169, 257 - 278.

Saito, K., (1971): Quasihomogene isolierte Singularitäten von Hyperflächen. Invent. Math. 14, 123 - 142.

Saito, K., (1973): Regularity of Gauss-Manin connection of a flat family of isolated singularities. In Quelques Journées Singulières. Paris: Ecole Polytechnique.

Saito, K., (1974): Einfach-elliptische Singularitäten. Invent. Math. 23, 289 - 325.

Saito, K., (1982): Primitive forms for a universal unfolding of a function with an isolated critical point. J. Fac. Sci. Univ. Tokyo Sect. IA Math. 28 775 - 792.

Scherk, J., (1980): On the monodromy theorem for isolated hypersurface singularities. Invent. Math. 58, 289 - 301.

Schlessinger, M., (1971): Rigidity of quotient singularities. Invent. Math. 14, 17 - 26.

Sebastiani, M., (1970): Preuve d'une conjecture de Brieskorn. Manuscripta Math. 2, 301 - 308.

Serre, J.-P., (1975): Algèbre Locale Multiplicités (3rd edition). Lecture Notes in Math. 11. Berlin etc.: Springer Verlag.

SGA 7', (1972): Groupes de Monodromie en Géométrie Algébrique, ed. A. Grothendieck. Lecture Notes in Math. 288. Berlin etc.: Springer Verlag.

SGA 7'', (1973): Groupes de Monodromie en Géométrie Algébrique, ed. P. Deligne et N. Katz. Lecture Notes in Math. 340. Berlin etc.: Springer Verlag.

Siu, Y.-T & Trautmann, G., (1971): Gap Sheaves and Extension of Coherent Analytic Subsheaves. Lecture Notes in Math. 172. Berlin etc.: Springer Verlag.

Slodowy, P., (1980): Simple Singularities and Simple Algebraic Groups. Lecture Notes in Math. 815. Berlin etc.: Springer Verlag.

Spanier, E., (1966): Algebraic Topology New York: McGraw-Hill.

Steenbrink, J. & Looijenga, E., (1983): A formula for μ-τ, to appear.

Teissier, B., (1972): Cycles évanescents, sections planes et conditions de Whitney. In Singularités à Cargèse, Astérisque 7-8, pp. 285 - 362. Soc. Math. de France.

Teissier, B., (1976): The hunting of invariants in the geometry of discriminants. In Real and Complex Singularities, Oslo 1976, ed. P. Holm, pp. 565 - 677. Alphen a/d Rijn: Sijthoff & Noordhoff.

Thom, R., (1964): Local topological properties of differentiable mappings. In Differential Analysis, pp. 191 - 202. London: Oxford U.P.

Thom, R., (1969): Ensembles et morphismes stratifiés. Bull. Amer. Math. Soc. 75, 240 - 284.

Wahl, J., (1981): Smoothings of normal surface singularities. Topology 20, 219 - 246.

Wahl, J.M., (1982): Derivations of negative weight and non-smoothability of certain singularities. Math. Ann. 258, 383 - 398.

Wahl, J.M., (1983): Derivations, automorphisms and deformations of quasi-homogeneous singularities. In Proc. Symp. Pure Math. Vol. XL: Singularities at Arcata. ed. P. Orlik, part 2, 613 - 624. Providence, R.I.: Am. Math. Soc.

Wajnryb, B., (1980): On the monodromy group of plane curve singularities. Math. Ann. 246, 141 - 154.

Wall, C.T.C., (1983): Classification of unimodal isolated singularities of complete intersections. In Proc. Symp. Pure Math. Vol. XL: Singularities at Arcata. ed. P. Orlik, part 2, 625 - 640. Providence, R.I.: Am. Math. Soc.

Whitehead, G.W., (1978): Elements of Homotopy Theory. Graduate Texts in Math. 61. Berlin etc.: Springer Verlag.

Whitney, H., (1965): Tangents to an analytic variety. Ann. of Math. 81, 469 - 549.

Whitney, H., (1972): Complex Analytic Varieties. Reading, Mass.: Addison Wesley.

Wolf, J., (1964): Differentiable fibre spaces and mappings compatible with Riemannian metrics. Mich. Math. J. 11, 65 - 70.

For the reader's convenience we separately list some books and conference/seminar proceedings pertaining to complex singularities.

BOOKS

Arnol'd, V.I.: Singularity theory (selected papers). L.M.S. Lecture Note Series 53 Cambridge: Cambridge University Press (1981).

Arnol'd, V.I., Varchenko, A.N. & Gusein - Zade, S.M.: Singularities of differentiable mappings (in Russian). Moscow: Nauka (1982).

Laufer, H.B.: Normal two-dimensional singularities. Ann. of Math. Studies 71 Princeton N.J.; Princeton University Press (1971).

Milnor, J.W., (1968). See the reference list.

Pham, F., (1979).

Slodowy, P., (1980).

SEMINARS AND CONFERENCE PROCEEDINGS

Proceedings of Liverpool Singularities I, II, Liverpool 1970. Ed. C.T.C. Wall, Lecture Notes in Mathematics 192, 209. Berlin, Heidelberg, New York: Springer (1971).

Singularités à Cargèse, 1972. Astérisque 7/8. Paris: Soc. Math. de France (1973).

Groupes de Monodromie en Géométrie Algébrique (SGA $7^{I,II}$), Bures 1969-1972. Lecture Notes in Mathematics 288, 340. Berlin, Heidelberg, New York: Springer (1972, 1973).

Several Complex Variables, Williamstown 1975. Ed. R.O. Wells jr. Proc. Symp. Pure Math. Vol. XXX. Providence R.I.: Am. Math. Soc.

Real and complex singularities, Oslo 1976. Proceedings of the Nordic Summer School/NAVF Symposium in Mathematics (P. Holm ed.). Alphen a/d Rijn: Sijthoff and Noordhoff.

Topological stability of smooth mappings, C.G. Gibson et al. Lecture Notes in Mathematics 552 Berlin, Heidelberg, New York: Springer (1976).

Séminaire sur les singularités des surfaces, Palaiseau 1976-77, ed. M. Demazure, H. Pinkham, B. Teissier. Lecture Notes in Mathematics 777. Berlin, Heidelberg, New York: Springer (1977).

Séminaire sur les singularités, Paris 1977-79, ed. D.T. Lê. Publ. Math. de l'Univ. Paris VII Paris (1980).

Complex Analysis of Singularities, Kyoto 1980. RIMS Kokyuroku 415. Kyoto University.

Singularities at Arcata (1981), ed. P. Orlik. Proc. Symp. Pure Math. Vol. XL. Providence R.I.: Am. Math. Soc. (1983).

INDEX OF NOTATIONS

$[(X,x)] \to [(Y,y)]$ (7.15)

$\mathrm{mod}[(X,x)]$ (7.15)

$\mathcal{C}is(n)$ (7.15)

$\mathcal{H}yp(n)$ (7.15)

$R^p f_* F$ (8.A)

Ω_f^p (8.4)

$\mathcal{H}^p$ (8.4)

DR (8.4)

$C_X^{\cdot}$ (8.4)

$\mathcal{O}(V)$ (8.9)

∇ (8.9)

$R(\nabla)$ (8.9)

ι_ξ (8.9)

L_ξ (8.9)

$\underset{\sim}{\theta}_f$ (8.9)

H_Z^q (8.13)

$\mathcal{H}_Z^q$ (8.13)

ω_f, ω_X (8.18)

$\tau'(X_0,x)$ (8.23)

F^p (8.25)

$M(f)$ (8.26)

ι_V (9.1)

SUBJECT INDEX